WITHDRAWN FROM
JAN 1 2 2010
USF LIBRARY

GROUP PSYCHOTHERAPY FROM THE SOUTHWEST

GROUP PSYCHOTHERAPY FROM THE SOUTHWEST

Edited by
MAX ROSENBAUM
Cooperating Editors
JOHN GLADFELTER
ALBERTO SERRANO

GORDON AND BREACH SCIENCE PUBLISHERS
LONDON NEW YORK PARIS

Copyright © 1974 by
    Gordon and Breach, Science Publishers Ltd.
    42 William IV Street
    London W.C.2.

Editorial office for the United States of America
    Gordon and Breach, Science Publishers, Inc.
    One Park Avenue
    New York, N.Y. 10016

Editorial office for France
    Gordon & Breach
    7–9 rue Emile Dubois
    Paris 75014

This title was first made available to the reading public in *Group Process* (Vol. 6, Number 1).

Library of Congress catalog card number 74–80279. ISBN 0 677 15790 8 (*cloth*). All rights reserved. No part of this book may be reproduced or utilized in any form or by any means, electronic or mechanical, including photo-copying, recording, or by any information storage or retrieval system, without permission in writing from the publishers. Printed in Great Britain by Clarke, Doble & Brendon Ltd.

CONTENTS

GROUP PSYCHOTHERAPY FROM THE SOUTHWEST

In Memoriam: William Sterling Bell	3
DAVID MENDELL The Southwestern Regional—A Community of Therapists	5
JOHN GLADFELTER The Liabilities of Individual Psychotherapy	21
MYRON F. WEINER, BARRY ROSSON and V. FRANK CODY Studies of Therapist and Patient Affective Self-disclosure	27
CHARLES W. ARCHIBALD, Jr. Group Psychotherapy with American Indians	43
ALBERT E. RIESTER and DINAH LEE TANNER The Leadership Laboratory—A Group Counseling intervention Model for Schools	49
LEWIS H. RICHMOND Observations on Private Practice and Community Clinic Adolescent Psychotherapy Groups	57
ROBERT L. BECK, PAT WIGGINS and IRVIN A. KRAFT What's in a Name? Name Assignment as a Pathological Function of Role Confusion in a Family	63
SIDNEY J. FIELDS Homosexuality: A Confused Trinity	73
MORTON KISSEN The Concept of Identification: An Evaluation of its Current Status and Significance for Group Psychotherapy	83
A. C. R. SKYNNER An Experiment in Group Consultation with the Staff of a Comprehensive School	99
HENRY P. POWERS, MARIA VICTORIA ACOSTA-RUA and RICHARD P. VORNBROCK Terminating an Open-ended Therapy Group	115
About the Authors	125

In memoriam:

These articles are presented as a tribute to the memory of William Sterling Bell, M.D.

DR. WILLIAM STERLING BELL died on Saturday, January 22, 1972. His death at age 46, of a myocardial infarction, was the termination of a long series of illnesses which had partially incapacitated him throughout life. Sterling, as he was called, was a plucky little fellow. His small stature and generally retiring ways belied the strength he had mustered to survive extensive surgery for regional ileitis while a teen-ager and to allow him to finish medical school and his subsequent psychiatric training. A native of Texas, Sterling was a graduate of the Duke University Medical School. His internship and residency in psychiatry were done at Southwestern Medical School, where he served on the clinical faculty until his death. His interest in group therapy dated back to his residency training. Sterling presented a paper entitled "Group Meetings on a Psychiatric Ward in a General Hospital" in 1957, which was subsequently published in the *Dallas Medical Journal*. It was probably the first presentation on group psychotherapy by a Dallas psychiatrist. Sterling had been president of both the Dallas Group Psychotherapy Society and the Southwestern Group Psychotherapy Society. Because of his active involvement in group therapy he was elected to Fellowship in the American Group Psychotherapy Association. He was an affiliate member of the Association for Group Psychoanalysis and Process.

Sterling's pluckiness and his willingness to fight carried him through his last and most difficult years with an air of dignity. He frequently suffered severe medical dysfunctions and toward the end of his life was forced to spend a prolonged period of time in a nursing home. It was as he appeared to be moving toward some degree of recovery that he died. Those who know will remember his devotion to others as people, to his profession, and in particular to his interest in the growth of group psychotherapy as a worthwhile, rational method of treatment.

MYRON WEINER, M.D.

The southwestern regional — a community of therapists †

DAVID MENDELL

The goals of the group therapist should be congruent with the goals of life. Carried over to his own organizational groups an experience in "living" should emerge from the work and service format and goal. The effectiveness of the group therapy mode in the "paratherapeutic" organization of our Southwestern regional organization has exemplified this in its history and development, as described here.

IN A GROUP established to achieve specific, task-oriented goals, the successful achievement of those goals leaves the group members with a feeling of satisfaction about themselves and each other and the group. They may then seek other goals to work toward, so they will not lose their emotionally rewarding relationship and joint effectiveness. At the conclusion of that effort, similarly, they will look for still other goals. In other words, the specific goals are also the means. They unite the group, making possible the constructive, satisfying human relationships that are, after survival, the ultimate goal in life, to my way of thinking. This paper traces the history of such a group, for the Southwestern Group Psychotherapy Society, which began as a regional professional organization, gradually became a community of therapists.

The inception of the Southwestern Group Psychotherapy Society dates back to June 10, 1956, when seven persons met in Galveston, Texas, to form a regional society of the American Group Psychotherapy Association (AGPA). This meeting took place at my instigation, for as a board member of the AGPA I had been given the mandate to form an affiliate society in the Southwest. I was

---

† The Southwestern Regional Group Psychotherapy Society, Affiliate of AGPA.

elected president, and we decided to hold scientific meetings with the following goals in mind:

1) Study and advancement of group psychotherapy to improve standards and practice.

2) Development of contact with others in the field of mental health and with those doing significant work with families and groups in order to contribute to the furthering and development of group therapy.

Formal affiliation with the national organization was the immediate administrative goal. The states to be included in the regional organization were Texas, Oklahoma, Arkansas, and New Mexico. From the very beginning as a psychiatrist with a psychoanalytic background and with some experience in handling groups, I attempted to establish group modes in administration and other functions of the society to provide a continuum with what was being taught and represented. That is to say, I hoped that the society would develop not only its subject but also its organization and being in the group psychotherapy mode. (I later coined the word *paratherapeutic* to refer to this learning-teaching-and-group-organizational model.) In such a group dynamic modality, learning has priority and therapy second place, in contrast to psychotherapy as such, where the priorities are reversed.

Due to the newness of the field and the frontier locale, most of the early participants in the regional organization were innovators and leaders in their own area. These areas, or subsystems of the regional society, were: Galveston, Dallas, and San Antonio, in Texas; Little Rock in Arkansas; Albuquerque, in New Mexico; and Oklahoma. The organizational and leadership thrust came initially from Houston, then successively from the various other areas.

A year after the society's founding, the first potentially disintegrative crisis occurred. The Galveston group, headed by Christopher Morris, feeling that the Houston group was not providing sufficient leadership, decided to break away and establish a local organization of its own. At the time, the regional society had just achieved affiliation with the national association, but had only a small number of members over the minimum requirement. This crisis was diverted by a new leadership thrust from Irving Kraft, a faculty member of the Baylor Medical School in Houston, a dedicated teacher who contributed a didactic and academic drive and interest to the society as well as the support of Baylor Medical School.

Southwestern began planning its first regional teaching institute—and a new crisis arose. It became evident that in the linkage of Southwestern and the Baylor Medical School two opposing modes of presentation had been joined—the experiential and the didactic. The polarity was resolved by choosing Alexander Wolf and Harris Peck, who represented the lecture and group dynamic extremes respectively, to be the leaders of the first two institutes.

In preparing for the first institute, another serious conflict over goal and direction developed when I, as the Society's link to the national organization, attempted to implement explicitly the national standards for recruitment of institute registrants. These were essentially the same as the qualifications for membership in the association. On the other hand, because of his didactic interests, Irving Kraft wanted to include all personnel interested in working with groups.† In part because the institute was to be held at Baylor, the more inclusive approach to registration prevailed. The resolving of these conflicts indicates the paratherapeutic function toward which the organization was struggling.

In January of 1959, eight months after the first institute with Dr. Wolf, the second was held with Dr. Peck, known to be a very dynamic and provocative leader. The thought had been that the second institute under his leadership would balance the more didactic, conservative first institute. However, the format established for the second institute, which was held at Baylor Medical School, remained in the didactic mode: formal, individual, one-hour lectures, followed by a half-hour question period and then dispersal into small groups for discussion of the lecture and other topics. (It later emerged that the lecture was rarely discussed in the small-group sessions.)

Dr. Peck was not content to go along with this format. He attempted to rechannel the course of the lectures, to involve the

---

† It is to be noted that his pioneering effort with nonprofessional therapists later became a recognized and widely accepted concept of paraprofessional personnel training.

lecture audience as a group, and to mobilize participants in the small-group sessions to make group decisions as to leaders and subjects. The resistances incidental to this action led to a showdown. Dr. Peck, caught between conflicting views in the society's leadership on the proper mode for the institute which he was directing, finally stopped the action, asked the two leaders Irv Kraft and myself to sit on two chairs on a stage by ourselves, presented us and the nature of the conflict in which he was involved to the audience, and left it to us to work out a solution to the impasse. The ensuing discussion occurred too late to salvage that particular institute, but it did help the audience to understand why the conference had been a fiasco, and it provided a primary learning experience for the local leaders.

The persistence of the didactic mode and the seeming resistance of the organization to the adoption of my proposed system led to my deciding to relinquish pursuit of this objective and stand back, letting the organization find its way as it would. To my amazement, the next president, Harold Winer, of Dallas, who had attended an NTL Laboratory in Bethel, Maine, began his term of office by pressing for the adoption of the modes that I had struggled to establish with so little apparent success.

In the fall and winter of 1959–60, a series of three bi-monthly one-day seminar meetings was held. These meetings, to which only members of the society were invited, included both the presentation of a scientific paper and discussion of the society's long-term plans. According to the minutes of the first such meeting, the majority felt that the objective of the series was to interchange information regarding group therapy and not to analyze the group's dynamics and interaction. "Personality involvement should be excluded. Personal analysis is not to be a function of the group." However, an informal gathering on the evening before the seminar meeting, that was characterized by frank and informal interaction, became a regular feature. A small core leadership group was beginning to emerge. The nucleus of a community was taking shape.

The next institute was held in Dallas in May of 1960, and was led by Florence Powdermaker. The format of the two-day meeting was three 45-minute theory presentations, each followed by small-group discussions lasting one and a half hours that were led by members of the society, Dr. Powdermaker expressed vigorous agreement with my idea of using group process to teach group psychotherapy.

With its fourth institute, held in San Antonio, in April 1961, Southwestern finally found its stride. Led by Hugh Mullan, this institute was one of the experiential process type—i.e. group psychotherapy was taught experientially. The institute itself lasted two days; however, the small-group discussion leaders and co-leaders (i.e. the society's members) met with Dr. Mullan on the days both preceding and following the institute, in what amounted to a training workshop and follow-up session. The interaction and learning experience provided by this institute so infused life into the San Antonio group of psychotherapists that it began one of the most vigorous growth spurts in the region. The other participants were equally galvanized, and Dr. Mullan, who had anticipated dividing his time equally between all the small groups of the institute got so caught up in the spirit and cohesiveness that he was unable to spend the time he had planned with his son, who accompanied him.

The following account drawn from notes written up at the time indicates the type of interaction that took place:

The most potent dynamic factor was the chairman's (Harold Winer) consistent referral of all significant decisions possible to the group of leaders which numbered about twenty and consisted of therapists of varying degrees of experience who desired a longer and more intensive contact with the guest speaker (four days) than that afforded by the teaching institute itself (two days). His first involvement of this group at the initial leaders' session was in making the selection of a leader and two co-leaders for each of the institute's seven small groups.

Then, after a short talk by the speaker, Hugh Mullan, the chairman, Harold Winer, proposed that those present divide up into two groups of 10 each for more effective interaction. This splitting up of the group had been planned in advance by the chairman and the speaker and had also been discussed with some others of the group, although they had had some doubts as to the feasibility. However, when the chairman made the suggestion at the actual meeting, he encountered resistance from the entire group, including the speaker. As he pressed his point, the group became provoked and attacked him. He then turned on one of the leaders, David Mendell, in anger for having urged him to retain administrative control of the meeting, while leaving the handling of the material to the speaker. In so doing, he unleashed much hostility evidently accumulated from as far back as the latter's tenure of office. This sudden blast, although it shocked the members of the group in various degrees, was quickly and animatedly disposed of. Thereupon, the speaker turned to me with the question, "What is there between *us*?"

No sooner had we undertaken this matter than several local participants, who were meeting with the regional society for the first time, and who had not been able to enter into active participation before this, burst out that they felt like outsiders and spectators at a private discussion. At this point the group came actively together in its entirety. A couple of latecomers, who had been sitting quietly behind the circle at the table, asked in on the front line. The group then remained together, long into the night.

The following day, as planned, the leader group met before the small-group sessions, then at lunch, and again after the small-group sessions. They expressed feelings about leaving their groups to join with the leader group for lunch. Some had already begun to sit with their small groups before joining the leader group.

At lunch the chairman reported, amid great hilarity, his attempt to visit three small groups and being thrown out, in effect. Later his feeling of administrative obligation to the organization as a whole, and his desire to share in the intimacy of small groups were less in conflict. During a subsequent session he visited several more groups without any reaction and the following session, he was able just to rest and enjoy it.

At the leader's session before the last of the small-group meetings, the matter of "closing" with the entire institute was brought up. A number of leaders reported curiosity in their small groups as to what went on in the leaders' group, and perhaps jealousy. Among various proposals was that of a testimonial to the preceding president and founder of the Southwestern regional society. He proposed instead that each leader sit in the large group with his group and briefly report on what had happened. Thus he could share his observations as well as his conflicts. If his small-group members did not agree as to what had transpired with them, they could then and there say so. There would thus be communication among the seven group leaders, as well as between each and his small group. This format was proposed as a means of maximal communication and sharing of the experience by way of closure.

There was some reluctance to this proposal. Under pressure of time the chairman decided on a compromise which would utilize about half of the group leaders. The proponent expressed disappointment.

After the small-group meetings, in which separation anxiety was reported as their groups broke up, lunch was served. The small groups sat with their leaders. A switch in arrangements then took place because of the noise in the dining room. The meeting was transferred to a large sitting-room and chairs were arranged to form a large two-tiered circle in an attempt to keep the selected speakers together for better interaction. Noteworthy about their presentation was the fact that one speaker felt the leaders' meetings had been boring and inconsequential and that his small-group meetings had been extremely rewarding. The following speaker, as he later realized, instead of reacting to this opinion, which contrasted with that of the leader group as a whole, went on with the observations he had come prepared to make.

The reaction from the whole institute was immediate and unfavorable. "We are being used by the leaders. Why are we still hanging around?" Some started to leave. An attempt was made to understand this falling apart. One interpretation offered was that it was a continuing to meet after a painful parting had been accepted; another, that its puprose had been inadequately communicated.

Finally the difference and some resentment of the dissenting opinion was expressed by the last speaker, whereupon the dissenters said, "Now we don't feel like going home any more." The meeting terminated with the feeling that only a partial resolution had been reached, but that there had been much going on that would give food for thought and learning.

When typing these notes, my secretary's impression was, "Looks like you all had a ball. That's the kind of meeting I like. I went to one last Friday night, and it was so dull I left."

After the San Antonio Institute, the paratherapeutic approach became the prevailing methodology of the Southwestern regional group. The developed goal was a learning experience for both instructors and registrants, with each group participating in psychotherapy-related sessions. The group sessions involving the instructors and the resource person were to give them firsthand experience of his leadership and provide him with feedback on *their* leadership.

The San Antonio institute format was modified in successive institutes as follows. The evening before the institute formally began, an informal meeting between the instructor group and the resource person was arranged, so that he would not enter the institute as an individual standing against the group. (Compare Martin Buber's I – It as against I – Thou). It began as a social evening with a buffet, but developed into a "free interchange" discussion that lasted far into the night. By the time the group broke up, many barriers had been broken and conflicts resolved not only between the resource person and the society, but among the members as well. This meeting, which would now be termed a mini-marathon, was a means of systems integration. The interchange continued at breakfast the following day.

The institute proper began with a brief introductory talk by the resource person. The instructors then met with small groups of registrants in a group-therapy type of session in which the registrants could have the experience of being group members and could also observe the modality of the local group instructor. This small-group interaction was the primary focus of the institute as the leaders' interaction and feedback with the resource person was for the faculty.

The instructors and resource person met together again over

lunch for feedback and a continued learning experience. Any anxieties, conflicts between co-leaders, discoveries, or noteworthy items were presented to the group as a whole and considered in group discussion, with the resource person serving as a leader and contributor until resolution and integration were achieved. In the afternoon, the instructors again went into session with their registrant subgroups, following which they regathered for dinner and a long evening with the resource person. The format was much the same for the next day. After the San Antonio Institute, the final feedback session was confined to the group leaders only.

The leaders in psychotherapy and the national organization who directed successive Southwestern Institutes† followed through and further developed the paratherapeutic process. Each added something from his own systems. Perhaps it was Martin Grotjahn who formulated his experience most aptly. He confessed to some anxiety when confronted with our unorthodox procedures. After the usual preliminary informal group interaction and exchange of information before the formalities began, he concluded that we were an "organization of brothers." The following morning he expressed the feeling that he had "passed through his adolescent initiation rites," and was anticipating his task with relish and without residual anxiety. (It might be noted that Dr. Grotjahn had at this time already begun his investigations into family and group therapy.) In view of his international stature in both the psychoanalytic and group movements, we felt that his enthusiastic interchange of information with us was particularly significant.

After Harold Winer, Irving Kraft served as president of the Southwestern society (1961–63), followed by Robert MacGregor of Galveston (1963–65). During this period, Paul V. Ledbetter, Jr., of Houston, performed the duties of secretary-treasurer with great

---

† Martin Grotjahn, Houston, April 1962; Sam R. Slavson, Dallas, Oct 1963; Jean Munzer, Little Rock, Oct 1964; Milton Berger, Houston, Nov 1965; Helen E. Durkin and Asya L. Kadis (co-leaders and spouses), San Antonio, Nov 1966; Clifford J. Sager, Dallas, Oct 1967; Carl Whitaker, Hot Springs, Oct 1968; Luis Feder, Corpus Christi, Oct 1969; Robert L. Goulding and Mary Edwards, Oklahoma City, Oct 1970; John J. O'Hearne and Lillian Plattner O'Hearne, Albuquerque, Oct 1971; Mansel Patterson, Oct 1972.

diligence and success. Others in the core leadership group besides those already named included W. Sterling Bell, Dorothy Carter, and John Gladfelter, of Dallas; Patricia Pearce, Marian Yeager and Diane Sheer, of Houston; Sidney J. Fields of Little Rock; Albert Serrano and Grace Jameson of Galveston; Roger Moon of San Antonio; and Richard B. Austin, Jr., of Corpus Christi; Lindell Cambier from Atlanta, Georgia, participated as well, until she organized the Southeastern society in her own area.

An episode that occurred at a local training institute illustrates the group dynamics among Southwestern's core members at this time. In February, 1962, Hugh Mullan came to Dallas to conduct a two-day institute, sponsored by the local group therapy society and psychiatry department of Southwestern Medical School. The first evening of the institute, W. Sterling Bell, a shy, gentle psychiatrist who had just acceded to the presidency of the local society, gave a buffet supper at his apartment for the instructor group and Dr. Mullan. I had been invited by Sterling, on behalf of the steering committee, to serve as one of the instructors. Harold Winer, the immediate past president of the local society, had also urged me to come, for he felt that I could be of help in resolving their local leadership problems. In his view, Sterling's leadership was somewhat autocratic and he was less process-oriented than others in the group.

At the buffet, however, I observed that as the guests were filling their plates and beginning to settle around the room, Harold sat himself in a large captain's chair that dominated all the other seating arrangements in the room. The instructors gradually clustered around this chair in a semicircle, facing him at the center. This activity ultimately polarized itself into a direct frontal assault on Harold taking him to task for this assertive behavior in not yielding to Sterling and in assuming for himself the prerogatives of office which he no longer held.

Without embellishing this occurrence, which took place in a period of about an hour and involved a very spirited exchange in which Harold valiantly tried to defend himself and his position, it is significant that it paralleled the actual dynamics within the local society. It was, in fact, difficult for Harold to give up his prerogatives and activities, and Sterling did not go after them. It was only after prodding and constant reminders from the other members that he gradually assumed his full responsibilities and duties.

After this discussion died down, Harold quietly and unostentatiously left the chair, which in the course of time was occupied by Sterling, and then Dr. Mullan, as the group settled into serious interchange late into the evening.

After the institute, Harold expressed in a letter his appreciation of my contribution in giving a sense of continuity with the overall commitment to group therapy and the regional society. He also reported that a great deal of the local anxiety, conflicts, and frustration had been resolved. He then mentioned the carry-over of a problem between him and me that still needed resolution.

Sterling, incidentally, was seriously organically ill; his life was maintained by cortisone injections and continuous medical supervision and hospitalization. He has since died. However, during this interval, he functioned well and happily in various positions, ultimately becoming president of the regional society, in 1968.† The medical members of our group felt that his health and vigor improved noticeably during this period, and it is our impression that his survival time was prolonged by his successful involvement in Southwestern. The presidency was offered to him not only as a function, a duty, to be effectively executed for the society, but also as a reward, an evidence of our love and appreciation for the effort he was making to sustain a high level of performance and participation.

Although satisfied with our effectiveness as a working group, some of us instructors began to express the felt needs for "something for ourselves." It was decided to expand the benefits of the developing pre-institute leadership workshop by inaugurating a retreat for the instructors that would be designed to improve their capacities but would involve no teaching responsibilities. The first such retreat was held at the Houston Yacht Club in April 1963, with Asya L. Kadis as guest leader. These retreats, or training workshops, which

---

† Others who have served as president of Southwestern are: Roger A. Moon (1966); Sidney J. Fields (1967); Paul V. Ledbetter, Jr., (1969); Albert Serrano (1970–72); and Richard Austin (1973).

alternated with the teaching institutes and became our most cherished and valued feature, took place in a quiet resort for a long weekend, and were limited to fifteen or so participants who were Southwestern members and practicing group therapists. The sessions met formally and informally from early in the morning until very late at night, and provided a laboratory for the study of interpersonal processes in relative isolation from distraction and contamination. It is an interesting fact that the now widely practiced marathon group appeared in other contexts about this time.

Several observations need to be made concerning this organizational development (the retreat group). First, it demonstrated anew that process rather than words was the fundamental communication model. Second, the retreats created a reservoir of good will among the retreat group members, which reached into their everyday practice and lives in the community. Third, there was no shying away from conflicts or disturbing personal problems. Various personal and professional conflicts between members, including immediate administrative power struggles, such as between a president and secretary, were brought out into the open and discussed. Personal problems of practice and family life also emerged and were dealt with and helped; these included some difficult and explosive situations, which carried over into the meetings as well (e.g. a depression, a disturbed, suspicious wife, and savage internecine rivalries within the group). In a few instances where these problems were not resolved or ameliorated in the usual course of events, the group and the visiting leader rallied around the troubled one—supporting him, making suggestions, or taking him to task. After he emerged from his ordeal, shaken, perhaps, by the experience, we often elected him president not too long thereafter as a gesture of confidence and support.

As time went by and new instructors were added, a subgroup developed in Southwestern made up of senior faculty, those who had participated in the society since its beginning and had taught in the regional institutes five or more years. These senior members took on the task of organizing psychotherapy training programs and setting up training standards. At this time some of Southwestern's personne and methodological models were taken up on the national scene.

In the mid 1960s in the national organization, the certification of group therapists was being discussed. Southwestern decided against doing this regionally but did establish a training model leading to the regional certification of senior faculty members. (This was done before training criteria for certification were developed on a national or local level.) Restricting ourselves to certifying only senior faculty members of the regional group enabled us to grant status and credentials on a basis that was justified locally, without having to assume excessive work or responsibility. Also, it was a conservative beginning in establishing certification models.

The next crisis that arose, in the mid sixties, illustrates the aptness of Southwestern's paratherapeutic designation. In that period a majority of its members, including instructors, who lived in remote sections of the region had very little contact with therapy organizations, analytic or otherwise. When Dr. Cornelius Beukenkamp, Jr., came to serve as retreat leader, in 1964, he made a proposal that actualized a potential crisis situation in the regional leadership. The senior faculty group had grown sufficiently large that it was beginning to exceed the number that could be comfortably accommodated in the cherished leadership group and its informal meetings with resource people at retreats and institutes. Two conflicting ways to meet this situation had been proposed. The first was to divide the leadership into two groups and have two resource persons at the retreats. However, this proposed solution raised the question of who would have to relinquish the primary group, and would one group have more status than the other.

The second proposal was a separatist one, that the senior faculty group close membership and cut loose from its organizational function, with autonomy and an identity of its own as a therapy group. A few other senior people and I objected to this proposal, because we saw the retreat groups not only as a training medium for the individual participants but also as the means for welding a therapeutic community by establishing better communication within the group. This in turn would facilitate both professional communication in our daily work and the planning and execution of the affairs of the society.

The separatist trend was reinforced by Dr. Beukenkamp, who

proposed that the leadership group become an independent psychotherapy group under his continued supervision. This, of course, was a major alteration of the original model, which had sought the greatest diversity possible in resource persons and had given first priority to information output and only secondarily accomplished therapy. Dr. Beukenkamp, a personal friend of mine, was hurt by my opposition to his plan; in implying that I was a drawback to the group, he clearly showed that he was in competition with me for its leadership. The faculty group expressed their love and appreciation of me, although aware of stubbornness of control on my part, but concluded, "We don't have to go along with him, and we don't." The majority of the faculty elected to go along with Dr. Beukenkamp's idea. The paratherapeutic aspect of the SWGPS faculty retreat thereby took second place to the therapeutic.

A partial split occurred in the membership. Those for whom therapy continued to be of major importance remained in the private therapy group, which Beukenkamp gradually transformed into such by charging his own fees and selecting his own additional members as older members dropped out. It developed into a separate subgroup with no intrinsic ties to the Southwestern society. Those older, more experienced members who were less in need of therapy gradually left his group and rejoined the paratherapeutic organizational model continuing in the annual institutes.

The original retreat modality was continued in modified form in the training committee meetings. These were held alternately with the institutes in a moderately isolated location for generally two days and had no outside leader. Group interaction had secondary priority to the planning of training, although the collegial atmosphere of the institutes continued.

At the ninth annual institute, held in November 1966, for the first time the spouses of the small-group leaders participated in a discussion group, Asya Kadis serving as leader. The intent was to let them experience the small-group interaction that was such an important part of their spouse's life.

Although the idea of such a group had been proposed several years earlier, I had not fully realized the spouses' feeling of isolation until the preceding year, when the institute was held in my hometown of Houston, and my wife, customarily most understanding and tolerant of my many absences from home to attend conferences, committee meetings, etc. this time became exceedingly angry. I then realized that she felt left out. The fact that she was excluded from the emotional closeness and sense of involvement within the leadership group (the result of our retreat and institute experiences together over the years) had become more apparent to her at this meeting because I was less physically absent from home yet obviously had my center of being somewhere else that weekend.

It is now clear in retrospect that the earliest manifestation of this sense of isolation was one wife following her husband to the first retreat and lingering at its periphery, because such was his intensity and dedication that she suspected him of having become romantically involved. The whole group, including the leader, helped this woman, who had become very depressed.

The spouses' group which had been most successful at the Southwestern's 1966 institute, seemed to be set to become a regular feature until some anxious repercussions came from AGPA. Because spouses were not eligible as group therapy trainees, the feeling in the national organization was that they should not take part in the institute as a separate group. Some other societies had attempted something similar and had not fared so well, perhaps because they were less well integrated. As a consequence, some of the national leaders were reluctant to come down for the purpose of leading such a group. Although an attempt made with a local leader did not go well, the separate spouses' group was not abandoned and couples-groups of instructors and wives were added.

Today the Southwestern regional society is in a state of equilibrium. The different parts of its organizational system are operating smoothly: new students enter and receive training qualifying them as members; instructors are upgraded and become senior faculty; the top echelon of leaders (in the main, the older members) serves as the training committee. The younger members are fully included in the delegation of duties. It is now an open system, taking in students and "putting out" senior faculty.

Those of us who got the regional society under way are no longer pivotal in its functioning. As a consequence we see each other less often than we would like. The less frequent contact notwithstanding, there is still a close, warm regard and refreshingly affectionate and easy conflict resolution between us socially and in matters that arise in professional relations and practice. Despite the distances in miles that have always separated us, and the spaces in time that now exist between our meetings together, and despite (or because of) the society's changing goals over the years, our sense of community continues.

Author's address: 1706 Medical Towers, Houston, Texas, 77025.

The liabilities of individual psychotherapy

JOHN GLADFELTER

The author discusses in a series of twenty points the liabilities of individual psychotherapy and explores briefly the limiting points in regard to this particular therapeutic approach. The article is written to evoke discussion and thinking in regard to psychotherapy which has been taken as standard tradition for many years.

IN MY TIME of doing individual and group psychotherapy I have encountered a number of liabilities of individual psychotherapy which I would like to enumerate for discussion. I am not making this list as a condemnation of individual psychotherapy but as a way of offering to other therapists a challenge and provocation of the status quo and an opportunity to reevaluate current thinking about therapeutic theories and approaches. I also present these views as a commentary on where I am as a therapist and as a possible signal for other growing therapists who may find themselves passing this same point in their growth.

1) I find that the development of the transference neurosis as a particular liability in individual psychotherapy. Although often sought for as a lever for growth and change and postulated as a necessary event in therapeutic process, transference has more often seemed an impediment for change and an opportunity for termination for the patient. When skillfully handled by well trained therapists, transference has been an important experience but too often has evolved into a neurotic process more difficult for the patient than his initial difficulties.

2) The rigid time structure of individual psychotherapy is both limiting and defeating for the efficient use of therapeutic time. Individuals at any particular moment in a therapy session may be able to use a time period of anywhere from five to twenty minutes but frequently find forty-five or fifty minutes burdensome and a matter of marking time for the end of the session.

3) The one to one encounter between therapist and patient is a limited experience resource for the individual growing in psychotherapy. Although I am a potent experience as a single human, I am only one and offer only one opportunity for experiencing humanness. For individuals with a sparsity of encounter with people and limited resources for obtaining such encounters, individual psychotherapy is a liability for such opportunities.

4) A single therapist is a limited and limiting social model in psychotherapy. Social learning and opportunities for interchange and social modelling and imitation are severely limited by the one to one situation. I can offer only a limited number of social opportunities where there is freedom to experiment and protection from social pressure in the conventional individual psychotherapy setting.

5) Individual psychotherapy creates, supports and enhances a secrecy myth about personal feelings, thoughts and fantasies that is unnecessary and potentially crippling to the individual in psychotherapy. I believe that the field of psychotherapy is at a point at which confidentiality and secrecy can be seen as the sabotage of human feelings in the service of so called help and protection. By the very nature of individual psychotherapy there is an endorsement of the myth of danger from others.

6) One to one psychotherapy introduces a regressive and artificial social world that is unique for overall adult human experience. There are no verbal, natural, one to one human encounters of forty to fifty minutes. The pressure of such encounters and the capacity of humans to learn to tolerate such experiences are a credit to the stamina of the human and are akin to the capacity of the human to tolerate urban living, traffic and boiler factories.

7) There is a clear lack of opportunity for potent social reinforcement of adaptive behavior and feelings in one to one psychotherapy. I as the therapist offer limited potency in this area and because of non-peer status and strict social proprieties of the conventional psychotherapy situation can seldom be seen as a social reinforcer.

8) There is emphasis on cognitive values and approaches in individual psychotherapy to the liability of action and affective

change. Insight and cognition have value in psychotherapy but by the nature of conventional one to one therapy the pressure for words and limitations for feeling and acting exaggerate the importance of thinking.

9) I think that there is a maintenance of social power stereotypes by the nature of individual psychotherapy. These stereotypes may have value but are usually well established in the individual who comes for psychotherapy and can be serious liabilities for the person who looks for ways to enhance his being and grow to a position of personal autonomy.

10) There is a fatigue and boredom experience for most people in individual psychotherapy which is deleterious to growth and change. The social isolation and the natural limitations of attention span contribute to these phenomena and not only limit but interfere with the learning and growing of the person.

11) The physical and social structure of individual psychotherapy create anxiety in a patient which does not enhance psychotherapeutic progress. Whether the psychotherapy be done face to face or side to side, the single social stimulus value of the therapist leads to a level of ongoing anxiety which consistently interferes with both cognitive and experiential learning for the patient. Often for the patient, there is the feeling that there is no place to escape which he maintains for the duration of therapy.

12) I believe that the cost of individual psychotherapy if being done by a competent and experienced psychotherapist is both limiting for many and prohibitive for some who would seek help. I think that this is true whether you consider agency or private practice as the major source of psychotherapy. When I observe the overall agency costs to society as balanced out against benefits therapeutically for the individual I find that the per person cost is not only as high or higher than on a private basis but that the limited potency of individual approaches makes individual psychotherapy a luxury which society can ill afford.

13) A liability of many individual therapeutic approaches is the inference that psychopathology resides within the individual. Further, this inference is based on a closed theoretical system. Such

an inferential system precludes either prolonged amelioration of symptoms or the notion of cure. With such an approach it is not surprising that psychotherapy takes many years and recurrence of symptoms and difficulties is frequent.

14) Individual therapies seldom allow the therapist access to a variety of models for treatment or psychological help. I know of few individual therapeutic approaches for instance that would consider utilizing a number of therapists or therapies.

15) Individual psychotherapy as generally practiced does not provide opportunity for the learning therapist to learn directly from his teacher or teachers either by direct observation or experience and involvement. Although supervision is common, there is a severe limitation in individual psychotherapy as to what the student may observe. I am not aware of any senior therapist who allows a student on a regular basis to observe in the therapy room the process of therapy.

16) There are few opportunities for the consideration of readiness on the part of the patient for therapeutic change in individual psychotherapy. The conventional time limits of fifty minutes combined with the enormous expense involved in three or four sessions per week make greater time opportunities economically unfeasible. I find it reasonable that if a patient is in a therapeutic situation for lengths of time of three hours or greater per week that his readiness and opportunity for therapeutic work will increase.

17) Individual psychotherapists have often argued that individual psychotherapy enhances and improves an individual's capacities to work and gain from group psychotherapy. My experience as a group therapist tells me that this is not only not so but that individual therapy at times interferes with and lengthens stay in psychotherapy and slows the person in making effective change.

18) Individual psychotherapy allows very limited opportunity for the individual to be aware of the full therapeutic potency of the therapist. The person in individual therapy often has only limited hearsay evidence for his choice of therapist and has little opportunity to know whether his therapist helps people to change and cure them.

19) There is a longstanding myth in individual psychotherapy that psychotherapy is difficult, painful, lengthy and doubtful in outcome. It is usually clear to the patient that he will not experience fun or joy in the process of change and would certainly not laugh or giggle. Such attitudes and positions about individual psychotherapy deter many people from choosing to work to make changes or choosing psychotherapy as a way of altering life patterns. The myth of pain created around individual psychotherapy is a liability to the process.

20) Traditional and popular notions about individual psychotherapy are that it takes a very long time, often many years and that even then, outcomes are problematic. Many people are discouraged by the time and the gloom surrounding psychotherapy and seek other ways of handling their difficulties.

The list of liabilities of individual psychotherapy is far from comprehensive but samples adequately the many difficulties I have encountered as a therapist. I do not believe that individual psychotherapy should be abandoned. As a group therapist I use a one to one approach in groups and believe that I am far more potent than when using only group process. I do believe that serious consideration should be given to the liabilities of individual psychotherapy in the development and planning of training in psychotherapy and in the development of mental health facilities.

Acknowledgement

I credit the Dallas TA 202 Seminar group for helpful comments and additions to this paper.

Author's address: 3524 Fairmount Street, Dallas, Texas 75219.

Studies of therapist and patient affective self-disclosure

MYRON F. WEINER, BARRY ROSSON and
V. FRANK CODY

The relationship between here-and-now self-disclosure of feelings by therapists and patients was explored in four short-term psychotherapy groups. It was postulated that the number of affective disclosures by patients would correlate positively with the number of affective self-disclosures by a pair of co-therapists. Criteria for disclosures were established, and scoring was done from videotape recordings. With the first two groups, therapists disclosed for five of the ten sessions, and did not disclose for five sessions. With the second two groups, therapists were disclosing in one group and not in the other. Analysis of the data revealed no statistically significant correlation between affective disclosure by therapists and affective disclosure by patients.

INTRODUCTION

IN recent years, a number of authors have suggested that a therapist's exposure of himself to patients in psychotherapy both enhances the therapeutic process and promotes self-disclosure by patients.

Carl Rogers (1961) states,

". . . the relationship which I have found helpful is characterized by a sort of transparency on my part, in which my real feelings are evident . . ."

Sidney Jourard (1964) writes,

. . . the therapist's openness serves gradually to relieve the patient's distrust, something which most patients bring with them into therapy. Still another outcome is that the therapist, by being open, by letting himself be as well as he lets the patient be, provides the patient with a role-model of growth—yielding interpersonal behavior with which he can identify.

Ian Alger (1969) proposes that the therapist,

> ... include his own behavior and personal reactions in the exchange he has with his patient, and the degree to which he can be direct and open in communicating this information will determine the freedom of his expression in the therapy. The corollary of this is that the patient will include his feelings and reactions in the same way, and indeed will be encouraged to be more free in this way himself by the example of the analyst.

There are few systematic studies in this area which relate to the actual psychotherapy of identified patients. Truax and Carkhuff (1965), in a study of 16 hospitalized schizophrenics in individual psychotherapy, on the basis of the study of 306 interview samplings, found the Pearson correlation between the average level of therapist transparency (willingness to be seen as a person) and the average level of patient self-disclosure to be $\cdot 43$ ($p < \cdot 05$). They concluded that the greater the therapist's transparency in the therapeutic encounter, the greater the patient's transparency or self-exploration throughout the course of therapy. Weigel and Warnath (1968), using groups, were unable to demonstrate differences between a group in which no such instruction had been given the leader, and a control group. Their conclusion was that their test instrument was not sufficiently sensitive.

Barrett-Lennard (1962) has measured "the counselor's psychological availability or willingness to be human," utilizing a questionnaire administered to clients which contained items such as,

"He will freely tell me his own thoughts and feelings when I want to know them."

"He is uncomfortable when I ask him something about himself."

"He is unwilling to tell me how he feels about me."

Barrett-Lennard found that the more experienced of his group of therapists were less "willing to be known" than the less experienced therapists. Willingness to be known on the part of the therapist showed no significant association with successful outcome of therapy.

Powell (1968), in 20-minute interviews with undergraduate students, found more emission of positive and negative self-reference when the interviewer matched interviewee's self-referring statements with statements about his own thoughts, feelings, or experiences about the pertinent topic.

HYPOTHESIS

We proposed to investigate, in our studies, one facet of the general thesis that honesty begets honesty and exposure begets exposure. We chose to examine the area of self-disclosure that seems to have received the most endorsement, namely, that if a pair of co-therapists in a psychotherapy group expose their own feelings in the here and now, there will be a corresponding exposure of here and now feelings by the patients.

STUDY No. 1

Patient sample

We attempted to enlist persons currently in individual outpatient psychotherapy, matched for self-disclosure on the basis of several screening tests. To obtain our patient sample, we contacted a number of therapists of different disciplines (psychiatry, clinical psychology, and psychiatric social work) and gave them the following information:

We are planning to conduct two short term therapy groups. We wish to offer patients in individual psychotherapy a brief group experience and at the same time gather teaching materials. We will accept as referrals any patient presently in individual psychotherapy.

The groups will meet in the Parkland Hospital† Clinic one hour weekly for ten sessions; they will begin in mid-January. A segment of each session will be videotaped. Participants will be asked to complete a questionnaire before and after the group to facilitate our evaluation of the groups. The fee will be set at half the fee for individual sessions, except for Parkland outpatients, who will pay their usual clinic fee per visit.

Of the 12 original applicants, we were able to include a total of nine patients in the study. Eight of the nine completed all ten sessions. One person dropped after five sessions. Table I gives some of the relevant data about each patient. No diagnosis was made. Our only criterion for selection was that each patient be in individual outpatient psychotherapy. Only one patient (J.S.) had a history of psychiatric hospitalization.

† Parkland Hospital is a city-county hospital facility.

TABLE I

Group A

J. M.	22,	WM	single	Once a week psychotherapy with	psychology graduate student
F. W.[a]	32,	WF	married	Once a week psychotherapy with	psychiatrist
B. T.[b]	32,	WF	married	Once a week psychotherapy with	psychiatrist
J. C.	30,	WF	married	Once a week psychotherapy with	resident psychiatrist
J.S.[c]	29,	WF	married	Once a month psychotherapy with	psychiatric social worker

[a]Private patient of M. W., terminated from individual therapy because of finances.
[b]Private patient of M. W.
[c]Terminated after five sessions.

Group B

P. H.	45,	WF	married	Every other week psychotherapy with	psychologist (Ph.D.)
B. M.	34,	WF	single	Once a week psychotherapy with	psychiatric social worker
R. P.	44,	WM	married	Twice a week psychotherapy with	resident psychiatrist
B. B.[a]	25,	WM	single	Once a week psychotherapy with	psychiatrist

[a]Private patient of M. W.

Matching of groups

We employed three tests for self-disclosure in an attempt to balance the groups for this variable. We used a 34-item modification of Jourard's self-disclosure scale to ascertain historical self-disclosure. (Jourard, 1971) In addition, each person was asked to make a list of areas he or she would be willing to discuss in a psychotherapy group.

Based on the Jourard instrument and the number of items indicated on the list, the 12 applicants were matched for self-disclosure, with six in each group. However, only nine of the 12 were in attendance. When we rechecked the balance of these two groups for self-disclosure after the first several sessions, we found that Group A had tested much higher than Group B.

Experimental design

Our initial plan had been for the co-therapists (M.W. and B.R.) to disclose here-and-now feelings in one group and not in the other. The group in which they were to expose (Group A) was decided by the toss of a coin. Because of the disparity of the self-disclosure scales between the two groups, we decided to reverse the procedure after the fifth session. Beginning with the sixth session, we discontinued our disclosures in Group A and began to disclose ourselves to Group B. We videotaped the last 30 minutes of most of the group sessions, and for scoring, utilized the third 15-minute segment of each group (i.e. from 30–45 minutes from the beginning of each group session) as the test segment. The test segments were scored independently by each of the authors according to the following criteria:

1) *A statement which requires no inference.* To be scored, a direct statement of feeling was required. Statements such as, "I'm sorry;" "I feel apologetic;" or "I feel defensive," were rejected as too vague. We felt they did not convey feelings directly, but rather only hinted at underlying guilt, remorse, or a sense of having to defend against an undefined feeling in one's self.

2) *A statement for which the patient assumes responsibility.* The **person**

who made the statement was required to "own" it. That is, we required that the person who declared a feeling assume responsibility for it. We rejected statements such as, "You make me feel sad;" "You are making me afraid;" and "It upsets me."

3) *A statement which is affective, rather than cognitive.* We did not score, "I am surprised," or "I am confused." We felt these were cognitive statements.

4) *A statement about a situation within the group in which the patient is directly involved, here and now.* We did not score feeling statements related to events outside of the group or that referred to past group sessions. We did not score statements which amounted to a diagnosis or a generalization, such as "I am an angry person." We did score sentences including "have been" when they referred to events in the same group session. For example: "I have been angry with you all this session, and I still am."

We did not score answers that were responses to direct questions which did not include a statement of feeling. For example:

    Q. "Are you angry?"
    A. "Yeah, a little bit."
    Q. 'Whom do you like better of the two girls?"
    A. 'Betsy."

5) We did score statements which followed the general formula:

    I   (feel)   (sad)   (with)   you.
        (am feeling)   (happy)   (toward)

We excluded statements which were essentially negations of feelings, such as, "I don't feel angry," or "I don't feel uncomfortable." We did score the phrase, "I don't like. . . ," because this seemed a statement of feeling rather than a negation of feeling. For example: "I don't like you," seems a clear statement of feeling.

Some of the self-disclosures by the co-therapists were: "I feel lonely;" "I feel uncomfortable with this discussion;" "I like you . . ." and "I'm getting frustrated." Some self-disclosures on the part of the patients were: "I feel good;" "I don't like it;" "I feel sad;" and "I feel guilty as hell."

Treatment technique

The psychotherapeutic technique employed by the co-therapists in these studies was highly confronting and placed great value on the expression of here-and-now feelings. We did not promise to cure our patients' emotional disorders in the 10 sessions allotted. Instead, we attempted during each patient's first group session to establish a limited therapeutic contract of which there would be some likelihood of success. For example, our contract with one woman was to help her feel less guilty about her personal need for her 9-year-old daughter, who was the only one of her 5 children she had been emotionally unable to give into the custody of her ex-husband, who was in better financial circumstances than she.

Scoring

A 15-minute segment (from 30–45 minutes) of each videotape was scored independently by each rater as indicated above. The raters each wrote out the phrases they scored as self-disclosures and noted

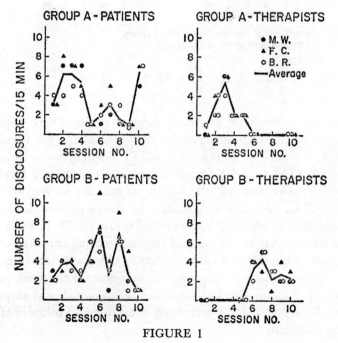

FIGURE 1

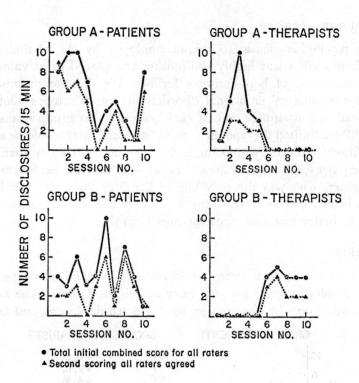

- Total initial combined score for all raters
- Second scoring all raters agreed

FIGURE 2

the footage on the tape at which they occurred. There appeared at times to be marked discrepancies between raters. (See Figure 1). However, for Group A (patients only), Kendall's Coefficient of Concordance, W was ·89, which is significant at the ·01 level. For Group B (patients only), W = ·75, significant at the ·02 level. Because of the variation in scoring, the raters pooled all the self-disclosing statements they had noted (Figure 1). Then, the pooled self-disclosures were reviewed, and the final scoring consisted only of statements acceptable to all three raters as self-disclosures (Figure 2, Table II). We observed that most of the discrepancies were related to lapses of attention on the part of the raters, which was surprising in view of the fact that we were only examining 15 minutes of tape from each session.

Results

1) *Prediction of self-disclosure.* Neither the Jourard questionnaire nor ("J") our list was an accurte apredictor of affective self-disclosure in the here-and-now (Table II). Spearman's rho rank correlation coefficients were: rho = ·004 for Jourard's scale, and rho = ·206 for our list of areas patients indicated themselves willing to expose.

2) *Relationship between therapist self-disclosure and patient self-disclosure.* There was positive correlation (Figure 1, 2 and Table II) between affective disclosure in the here-and-now by therapists and here-and-now affective self-disclosure by patients in our short term therapy groups. We were unable to apply statistical measures to this data because of the premature termination of J.S., the most self-disclosing member of Group A. Simply dropping J.S.'s score from the data reverses the conclusion for Group A. Had we added in an expected score for the last five session for her, the conclusion would have been the same. We felt, however, that her decision to drop from the group was a decision against further disclosure.

TABLE II

GROUP	SESSION No.										TOTAL	"J"		
"A"	1	2	3	4	5	6	7	8	9	10		Pre	Post	List
J.M.	1	0	1	1	0	0	4	1	0	1	9	3	7	4
F.W.	0	AB	3	0	0	2	0	0	0	0	5	6	10	3
B.T.	0	1	2	AB	0	0	0	0	1	1	5	24	23	8
J.C.	0	1	1	0	0	0	AB	0	0	4	6	Totals		15	10	9
J.S.	8	4	AB	4	0	DROPPED					16	1-5	6-10	23	—	15
Total	9	6	7	5	0	2	4	1	1	6	41	27	14	
Leaders														
M.W.	1	2	2	1	1	0	0	0	0	0	7	Totals		
B.R.	0	1	1	1	1	0	0	0	0	0	4	1-5	6-10	
Total	1	3	3	2	2	0	0	0	0	0	11	11	0	
												"J"		
"B"	1	2	3	4	5	6	7	8	9	10	Total	Pre	Post	List
P.H.	AB	0	0	0	0	1	0	5	2	0	8	0	1	5
B.M.	0	0	0	0	3	0	0	0	0	0	3	13	16	4
R.P.	2	1	3	0	0	5	1	1	0	0	13	Totals		9	14	4
B.B.	0	1	0	0	0	0	0	0	1	1	3	1-5	6-10	6	11	7
Total	2	2	3	0	3	6	1	6	3	1	27	10	17	
Leaders														
M.W.	0	0	0	0	0	3	2	2	2	2	11	Totals		
B.R.	0	0	0	0	0	0	2	0	0	0	2	1-5	6-10	
Total	0	0	0	0	0	3	4	2	2	2	13	0	13	

Discussion

There are obviously many uncontrolled variables in this study. Our groups were smaller than what we had hoped for. J.S., the most self-disclosing member of Group A, dropped after five sessions, and this may have contributed to the lessening of self-disclosures in the latter five sessions. There was much feeling about her termination. The group members felt unable to cope effectively with her tendency to feel maligned and criticized, but undoubtedly experienced guilt, sorrow, and a sense of helplessness in response to her termination, as well as some relief.

Our procedure allowed us to use each group as its own control, but the groups were not strictly comparable in that we made our self-disclosures to one group at a different point in its history than we did with the other group.

There was a discrepancy in self-disclosure between the co-therapists. This may have created some ambiguity about the desirability of, or necessity for self-disclosure.

Another factor for which we could not account were the reactions attributable to a blend of personalities unique to every group. Both groups expressed a strong wish to continue. but only in Group A was there a really intense emotional expression of this wish, and of a sense of loss at termination. Group B tended to deal with termination in a much more intellectual manner.

STUDY No. 2

The patient sample

The co-therapists, (F.C. and M.W.) in contrast with our former study, had prior contact with only one member of this study population. In our initial study group, one of the authors (M.W.) had made personal telephone contact with virtually all of the group participants, and 3 of the patients in the study were or had been in individual therapy with M.W. for varying periods of time. We required much more cooperation from our first study group than the second. In our earlier group study, we required that a number of complicated questionnaires be completed prior to the beginning of treatment. We did not attempt to match our second groups in terms of predicted self-disclosure because we had found no correlation between the results of predictive questionnaires and here-and-now affective self-disclosure in our first study group. (Weiner, 1971) Patients were randomly assigned to two groups. We decided on the flip of a coin the group (Group A) in which the co-therapists would be self-disclosing.

Referrals were obtained from the Outpatient Psychiatric Division of Parkland Memorial Hospital. The only prerequisites for referral to our study were that the patient be over 21 (to be able sign to a consent for videotaping our sessions), that the patient had not recently suffered a psychotic decompensation, and that the referring mental health professional felt a short-term group experience might be of value to the patient as an individual. All prospective members were told that they would be meeting for 10 one-hour sessions at weekly intervals, and that some portion of each session would be videotaped for teaching and research purposes. We did not allow the patients to see the videotapes and explained this by stating that it was not part of our research design, and that we would have to rely on each other's honesty rather than the T.V. equipment for a picture of each person as he came across to others. To our knowledge, only one of our patients was in any sort of concomitant therapy. Our patient population ranged in age from 21 to 54 with an average age of 28. Our sole male participant was Negro. (See Table III).

Collection of data

Our measurement of here-and-now affective self-disclosure was based on randomly-timed 15 minute segments of the first eight sessions of each group. The co-therapists were unaware, except for two sessions during which there were technical difficulties, of the portion of each session which was videotaped. We utilized only one rater (M.W.). One group lasted 10 sessions, the other only 8.

TABLE III

Group A	MS[a]	\	\	\	Session No.	\	\	\	\	\	\	Total No. of sessions attended	Av. attendance per session
		1	2	3	4	5	6	7	8	9	10		
H.E. (51 WF)	(W)	X	X				X	X	X			5	
M.T. (27 WF)	(M)	X	X		X			X				4	
R.M. (24 LAF)	(S)	X	X									2	
A.W. (21 NM)	(S)	X	X	X								3	1·7/10 sessions
P.M. (35 WF)	(D)	X				X						1	
P.M. (35 WF)	(D)		X	X			X					3	2·05/8 sessions
Total Attendance per session		5	5	2	1	1	2	2	1				

Group B	MS[a]	\	\	\	Session No.	\	\	\	\	\	\	Total No. of sessions attended	Av. attendance per session
		1	2	3	4	5	6	7	8	9	10		
T.F. (28 WF)	(D)	X	X	X								3	
C.F. (24 WF)	(M)	X	X	X	X	X	X	X	X	X	X	9	
E.R. (22 WF)	(S)	X	X	X	X							4	3·7/10 sessions
D.H. (21 WF)	(M)	X	X	X	X	X	X	X	X			8	
M.B. (21 WF)	(S)			X		X	X					3	4·0/8 sessions
Total Attendance per session		4	3	5	3	3	3	2	2	2	1		

[a] MS = marital status

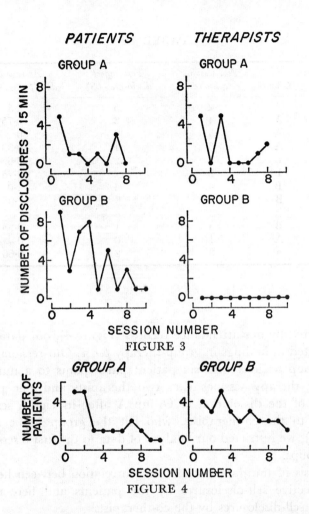

FIGURE 3

FIGURE 4

Data

The correlation between total here-and-now affective self-disclosures when therapists were self-disclosing is presented in Figure 3, as is the correlation between affective self-disclosures by patients in the circumstance of therapist non-disclosure. Figure 4 shows the attendance record for both groups. Table IV assembles our data in numerical fashion, and it is from this data that our statistical analysis was made of the correlation between here-and-now affective disclosures on the part of group leaders and here-and-now affective disclosure by group members.

TABLE IV

Name	Group	No. sessions attended	Total affect disclosure (8S)	Average affect disc. per session (8S)
R.M.	A	2	5	2·5
M.T.	A	4	3	0·75
H.E.	A	5	0	0
A.W.	A	3	1	0·33
T.F.	B	3	2	0·66
D.H.	B	8	22	2·75
E.R.	B	4	4	1·0
C.F.	B	9	8	1·0
B.M.	B	3	1	0·33
P.A.	A	1	0	0
P.M.	A	3	2	0·66

Results

As reflected by our attendance records (Figure 4), our data cannot be reported as findings in group therapy *per se*. Our sessions ranged from group sessions with six patient participants to a number of multiple therapy sessions (i.e. two therapists and one patient). Because of the dissolution of Group A after its eighth session (a surprise to the co-therapists, who felt the group to be strongly cohesive), we restricted our analysis of data to the first 8 sessions for both groups.

Analysis of the data revealed no correlation between here-and-now affective self-disclosures by the patients and here-and-now affective self-disclosures by the co-therapists.

There was some relationship, but *not* statistically significant, between the number of persons present at a group session and the amount of here-and-now affective disclosures. As attendance dropped in group A, it became more difficult for the co-leaders themselves to disclose here-and-now feelings. Attendance held up better (average of 4 per session versus 2), and patient self-disclosure was greater, when therapists were non-disclosing (total of 37 affectiv- disclosures versus 11). but neither finding can be statistically valie dated because of the sample size.

Conclusions

These studies raise the question of the feasibility of manipulating a variable such as self-disclosure, which is strongly related to situational and unconscious intrapsychic factors, and not a highly flexible behavior that can readily be consciously controlled, such as the volume of one's voice or one's speed of walking.

The process of psychotherapy, if conducted at a level which leads to greater self-understanding, does involve self-disclosure on the part of both patient and therapist. In dealing with the issue of self-disclosure in psychotherapy and in ordinary interpersonal relationships, the basic question is not its presence or absence, but the how, why, when, and where. Restated simply, one must ask, "self-disclosure to what end?" If self-disclosure is good for patients in psychotherapy, what means of eliciting disclosure are valid? And when? (Weiner, 1972)

Our conclusion is that disclosure by the therapist is not necessarily facilitating of patient self-disclosure. Our further conclusion is that research based on determining the sheer number of self-disclosures by therapist and patient over a given time segment may be of less value than a study of the nature and the timing of the disclosure by the therapist as related to its impact on the patient. Meltzoff and Kornreich (1970) suggest that future research on the therapeutic relationship will need to take into account the interaction between the patient and therapist rather than therapist-offered conditions alone. Our difficulty in being self-disclosing as attendance dropped in Group A certainly substantiates this point of view.

Summation

In contrast to laboratory studies, in which interviewer self-disclosure mitigates in favor of interviewee self-disclosure, there is little evidence from clinical studies that self-disclosure by the psychotherapist is strongly associated with increased self-disclosure by patients.

References

Alger, I. (1969): Freedom in Analytic Therapy. *Curr. Psychiat. Ther.*, 9, 73.
Barrett-Lennard, G. T. (1962): Dimensions of therapist response as causal factors in therapy change. *Psychological Monographs*, Vol. 76, whole No. 562.
Jourard, S. M. (1964): *The Transparent Self*. New York: Van Nostrand.
Jourard, S. M. (1971): *Self-Disclosure: An Experimental Analysis of the Transparent Self*, Wiley-Interscience, New York.
Meltzoff, J. and Kornreich, M. (1970): *Research in Psychotherapy*, Atherton, New York.
Powell, W. J. (1968): Differential effectiveness of interviewers' interventions in an experimental interview, *J. Counsel. Psychol.*, 32, 210.
Rogers, C. R. (1961): *On Becoming A Person*. Boston: Houghton-Mifflin.
Truax, C. B. and Carkhuff, R. R. (1965): Client and Therapist Transparency in the Psychotherapeutic Encounter. *J. Counsel. Psychol.*, 12, 3.
Weigel, R. G. and Warnath, C. F. (1968): The Effects of Group Therapy on Reported Self-Disclosure. *Int. J. Group Psychother.*, 18, 31.
Weiner, M. F. (1972): Self-Exposure by the Therapist as a Therapeutic Technique. *Am. J. Psychother.* 26, 42
Weiner, M. F., Rosson, B. and Cody, V. F. (1971): Prediction of affect disclosure in group psychotherapy, unpublished manuscript, University of Texas Southwestern Medical School at Dallas.

Authors' addresses: Myron F. Weiner, Clinical Associate Professor of Psychiatry The University of Texas Southwestern Medical School, Dallas, Texas 75235. Barry Rosson, M.D. Menninger Foundation, Topeka, Kansas. V. Frank Cody, M.D., Instructor in Psychiatry The University of Texas Southwestern Medical School Dallas, Texas 75235.

Group psychotherapy with American Indians

CHARLES W. ARCHIBALD, Jr.

Unique communication characteristics of American Indians modifies the normal expectancies of the group process, requiring specific shifts in technique for even the experienced therapist. The Indian's comfort with protracted silence, an extreme sense of personal privacy coupled with a guardedness for fear that cultural secrets may be revealed and lost, and the acceptance of individual difference being a right, makes the non-Indians quest for the "why" of behavior antipathetic to the Indians cultural orientation. In a search for an approach that will utilize the strong sense of oneness with all Indians that each feels, an attempt has been made to utilize the individual Indian's sense of striving to remain in harmony with his environment. Through drawing him out regarding those thoughts and actions which cause abnormal stress to him and those about him, the group members have achieved an awareness of a group process that is compatible with their Indian ways.

BEGINNING in the 1950's with attempts to help the Indians express their feelings regarding adjustment to urban living produced by the Bureau of Indian Affairs Relocation Program, behavioral scientists skilled in the techniques of group psychotherapy have been trying to utilize group therapy techniques with Indians. They have offered young Indians in the BIA boarding schools a chance to speak out regarding the uses and abuses of the educational settings often far removed from their homes. The acting out adolescent has been a popular referral from the reservation school staffs, public or parochial, especially the disturbing boy in class and the girl whose social behavior is identified as disruptive to the peer group. The next most common use of group techniques for personal growth has been with the staffs that serve Indians, usually a mixture of Indians and non-Indians. There have been Indian couple groups, although most commonly the identified problem was the husband's drinking. There has been little experience with the heterogeneous group, probably the most common composition of non-Indians in treatment.

While group treatment has been proudly listed by staffs of settings with a psychosocial remedial or rehabilitative function, little has been written of the success of these ventures. Due to the author's interest in improving service to the Indians of the Southwest, informal interviews of professional helping persons in the western states of California, Washington, Nevada, Minnesota, Kansas, Arizona, Colorado, New Mexico, Utah, and Oklahoma were conducted during conferences on Indian psychosocial issues, such as Alcoholism, Drug Abuse, Suicide, Law and Order, and Indian Education. Initially these skilled practitioners expressed generalized frustration as they reviewed the attempts to achieve the type of group cohesiveness through group process that they had experienced with non-Indian groups. With support they were able to admit that the fault was not with the Indians, but with their own inadequacy to sense "where the Indians were". The few that claimed to be making progress seemed to divide themselves between (1) those who supported the Indians' willingness to discuss the persecution of their people, and (2) those who would put forth standards of the dominant culture and then encourage discussion. I have fallen into both of these traps, and would identify the first as a defense which has been found very hard to shift from once established, and the second as invitation to intellectualization that may evolve into something more personal through the efforts of the leader and/or strong group members. No one claimed to have a successful psychotherapy group with more than a 50% Indian composition.

In seeking the common characteristics of the thirty-two tribes represented in this sampling of reservation-based Indians, three traits seemed particularly worthy of noting. The first all agreed on silence as a defense. While common in group psychotherapy among non-Indians, with the first Americans it has been developed to a high degree. It is suspected that the respectful silence of youth in the presence of their elders is a secondary influence to the suppression of the openness of the Indian child by significant, usually non-Indian school teachers and religious spokesman, who have belittled their efforts to express themselves in their seeming naivete. They have learned to seal over their deep pain, and the more stable are able to draw into themselves for strength in knowing who they are in order to remain in harmony with their personal world. To the outsider they show the bland countenance which hides the unspoken, "I am no fool, and I will not give you words as ammunition to turn against me in your quest to make me appear a fool"! This is the first challenge to the non-Indian therapist.

Second is the group solidarity, which has several components. On a superficial, but strongly felt level, there is the insistence that approaches are done "in the Indian way," accepted as right by other Indians without explanation of the particulars. Invariably these are based on "common sense" and "experience", which are not seen as unique to the individual. Then there is the fear that to express details of approach and its reasons may unwittingly reveal an Indian belief. Any attempt to reason, "Tell me, so that I may better understand you through your orientation," is met by the patient insistence that a non-Indian could not possibly understand; masking a fear that they will be belittled, and their revelations misused to the detriment of their people. Finally there is the norm of the group itself that disapproves of a member setting himself apart in either a claim of greater strength or an admission of weakness. The unspoken fear in the latter case is a group extension of the fear of the individual non-Indian prior to the first time he expressed weakness, "If they know this about me they will turn away." Invariably just the opposite happens, and such occasions are milestones in building group solidarity during its early stages. Not so with the Indian group; the taboos are strong enough for the individual who sets himself apart by "breaking the code," that his worst fears of rejection may be realized. True, this characteristic is not exclusive to Indians—I have experienced it with the criminal sub-culture of the heroin addict; but it is a significant challenge to the non-Indian therapist.

The final characteristic that should be mentioned as a near universal, although identified in different ways by the professionals interviewed, was the Indians' insistence that a major difference between the Indian and his non-Indian counterpart is the latter's tendency to ask "Why?" The Indian claims to accept things as they are, both for himself and those about him. Paradoxically he does not hesitate in the least to give advice, explaining when he says, "If I were you . . .," he is quite literally putting himself in the other person's position, an extension of the quite acceptable principle of "starting where the other person is." The fallacy of this reasoning need not be explained to the professional reader.

Thirteen months ago a new group was formed, made up of the staff of an alcoholism treatment center. The group met only twice monthly, for two to three hour sessions. There have been twenty-three sessions to date and are expected to continue indefinitely. Eighty-five per cent of the group have been reformed alcoholics, and ninety per cent have been Indians. Nineteen people have participated, with an average group attendance of ten. Everyone attends from Director to Secretary, to Cook; justified on the basis that all have contact with the client group, and need to know how and why they come across to others.

This particular group was chosen because they had expressed a goal of developing an alcoholism treatment program utilizing traditional Indian ways, viewing alcohol as an adulterate to their goal of harmony between the individual and his environment. Two members had been introduced to sensitivity group experiences as part of a year's university training as alcoholism counselors, and offered support to the therapist when other staff members questioned the worth of looking inward in order to function better outward.

Although problems "brought from home" were identified in examining the contribution of the individual member to the harmony or disharmony of the group (ongoing treatment program milieu), the focus was on the interaction of the members present in the group for a given session. At first this came slowly, and many attempts were made to break away with intellectualizations of, "Indians feel that . . .," but these were consistently countered with, "What effect do you think your words and actions have had on the program?" By the fifth session other members were tentatively supplementing the impressions of the person on the hotseat, but with surprisingly little openly expressed hostility. Now when they said, "I can understand how he felt," it seemed believable.

No longer are personal differences being permitted to pile up, only to explode a month later to shatter all. The semi-monthly sessions are being supplemented by daily mini-group meetings for up to half an hour at the close of the daily staff meeting, led by the co-therapist who has emerged. Incidentally, she is an Indian, but from a tribe 900 miles away from the host reservation where 80% of the staff are enrolled members.

There are, of course, many other approaches that are being tried with Indian people utilizing group treatment techniques. They are a communal people, and group acceptance is life itself, while ostracism is the ultimate punishment for a reservation-based Indian. The sins of General Custer and his kind fall heavily on his kinsmen as they attempt to relate without a perpetuation of an imposition of standards of behavior upon the Indians by the dominant culture. This need never be an expectation placed on the group therapist, whether he is working with drop-outs, delinquents, or drug abusers; despite the expectations of those who invite or permit the therapist in. The Indian, like anyone else, will do that which makes sense to him; not what he is told to do. The duty of the therapist and the group is to guide him in his search for understanding and direction compatible with his Indian ways. For the non-Indian this requires an abnormal amount of patience, acceptance, and trust in a group process that has several unique components. If he "Keeps the faith," he will find in the heart of the Indian a richness of spirit that has permitted a people to face adversity for almost five hundred years of dealing with the discontented Europeans, and to draw strength from all that is about them to help them survive. Your efforts can be a part of that strength if you are able to walk a ways in their moccasins.

Author's address: Dept. of Psychiatry, University of New Mexico Medical School, Albuquerque, N.M. 87106

The leadership laboratory
A group counseling intervention model for schools

ALBERT E. RIESTER and DINAH LEE TANNER

The article describes a group counseling program which was conducted in an elementary public school for children with a variety of inter-personal and school adjustment problems. The article illustrates the advantages in having a mental health specialist from a child psychiatric facility conduct the group as co-leader with the school counselor. The outcome of this group counseling activity helped the teachers become more sensitive to the affective needs of the children and illustrated the advantages of team work in designing programs to stimulate the child's overall cognitive and emotional development.

LEADERSHIP LEARNING LABORATORY is a group counseling intervention approach occurring during the school day, conducted on campus; it was an attempt to give elementary school children an opportunity to identify ways to work on both personal and school adjustment problems. The "leadermanship," as students frequently called it, resulted from an elementary school counselor's awareness that during the school day children need time to resolve a variety of school and interpersonal problems. The counselor and faculty also felt that some students had school adjustment problems requiring counseling of some form. After discussing this general student need with teachers and the principal, the counselor contacted the Community Guidance Center because of their interest in school based service and consultation programs. After a number of planning meetings, the principal, faculty, counselor and Community Guidance Center consultants agreed that the counselor and psychologist from the Guidance Center would offer a counseling group to focus on the affective needs of approximately 12 children and that they would be co-leaders of this group. This first group would be a pilot project, the results of which would be evaluated for possible incorporation of this intervention approach into other schools in the district. Also, the group would give the counselor an opportunity to develop skills in group counseling and the Guidance Center could learn more about the school system.

During the initial planning sessions, careful steps were taken to provide frequent opportunuties for teachers and administrators to become aware of the goals of the group, as well as open discussions of the issues which develop as a result of the introduction of a new program.

SELECTION OF STUDENTS

Students nominated by teachers were selected for the group by the co-leaders. One stipulation to nomination was that the nominee have some school adjustment problem which was interfering with his social and/or academic functioning. The co-leaders selected twelve out of 35 children on the following criteria: (1) They were representative of the school population; (2) The group would be co-educational; (3) The group would be representative of the full range of academic achievement levels. (For example, students who had excellent grades were included if they had interpersonal problems with peers and/or adults); (4) There was a clear indication that the child did not have emotional problems or language learning disabilities requiring treatment by established intervention modalities such as speech therapy or referral to an outpatient psychiatric facility.

After the group selection, the counselor reviewed the criteria with the teachers so they understood why other candidates were not picked for the group—an important step because many of the teachers had hoped that the children they had nominated would receive the benefits of group counseling.

The final group included representatives from four peer sub-cultures in the elementary school social system. These groups could best be described as "Sneaky Sam", "Sorority Sue", "Mad Mafia", and "Passive Pete"; all generally representative of the school's student population.

Parents were contacted personally by the counselor who made home visits to explain the purpose to parents and child and to clarify any questions and/or mis-understandings. The possibility of learning leadership skills and more effective interpersonal skills with both adults and peers was stressed. Written parental permission was obtained which included authorization for co-leaders to take the children on field trips off campus during school hours. Full parental support for the new group was thus achieved. Children and parents alike viewed this opportunity to participate in this group as a privilege and unique educational opportunity. A convenient meeting place and time in the school was arranged with each session lasting 90 minutes and occurring weekly.

GOALS OF THE GROUP

The following goals set by the group co-leaders were communicated to parents, faculty, and administration.

1) The group experience would provide an opportunity to learn problem solving skills in conflict situations. Example: Bill was concerned about his grades and the group outlined some ways to request help from the teachers.

2) The co-leaders would focus on positive behavior exhibited by all children. Example: A withdrawn child is given recognition by the group leaders, allowing other children to verbalize their feelings to a person who often is not recognized by others. A child who was pushy and aggressive is given social reinforcement when he is more sensitive to the needs of the other members of the group.

3) The co-leaders would follow the rules and regulations of the school system. Example: John decided to walk away from the group in the middle of a session. He had the choice of either returning to the classroom or remaining with the group. Roaming on the campus would not be tolerated by the group.

4) The co-leaders would provide the children with ideas and

insights concerning the resolution of problems which occur in everyday living. The co-leaders of the group functioned as consultants and resource people to the children when they presented a variety of family problems and interpersonal problems with teachers and administrators. Example: The co-leaders would not sit back and withhold information which would permit the development of a more successful group. Field trip suggestions were often furnished by the co-leaders and they offered new approaches in relating to their teachers and parents.

5) The co-leaders were to be open and honest with each other in front of the children in resolving their own misunderstandings and communication problems, thereby becoming models for conflict resolution. Example: When confronted with his frequent absences from the meetings, the psychologist shared some of his own problems in trying to meet his professional obligations, and he also decided to rededicate himself to the group.

6) The group experience would expand the child's self awareness of how their behavior is perceived by others. Example: Sue was confronted about her aggressive and pushy way of dealing with the other group members. When she was in a leadership role, she would always "railroad" the others into doing what she wanted. By role playing and guidance, Sue learned to obtain opinions from the group members and follow more democratic decision making procedures.

THE FIRST MEETING

At the initial group meeting, the co-leaders stressed the group was an opportunity to develop leadership skills, something elementary children can understand and discuss. Adult co-leaders identified the leadership behavior present in the group which was essential in permitting a group to function. Initially, the children were asked if anyone wanted to volunteer for the leadership role in the group to see what it would be like. This child would then switch chairs with the adult co-leader and be responsible for discussion and direction of the group for a few moments. When this occurred, his leadership strengths were identified as were his deficiencies; it was noted that many individuals have limited knowledge of how to assume a leadership role and the group was a laboratory to learn more effective group work skills. Confrontations were designed to be psychologically nonthreatening, as children are extremely sensitive to negative feedback about their behavior. When the co-leaders used the nonthreatening, constructive and positive approach in pointing out leadership deficiencies, the children likewise began to help other group members in a supportive way.

The first session also focused on routine administrative and management problems such as meeting time and stipulation that school rules would apply during group time. To further insure a successful group, the co-leaders would point out that this could be an opportunity to go on field trips provided necessary planning was accomplished and permission obtained. The co-leaders during the first session responded to the enthusiasm of the group and agreed to go on a field trip the following week. Adults were resource people assisting with planning and were active in formulating organizational details for the field trip. This illustrates the general philosophy that children require adult guidance with co-leaders sharing important information which would insure group success. Co-leaders would point out more efficient, effective modes of planning and organization.

THE FIELD TRIPS

During a field trip, the co-leaders observed behavior ranging from leaving the group to extreme shyness. These behaviors were subsequently discussed and the children began confronting each other about their actions. Group cohesiveness slowly emerged as the group formulated their own rules and the children became more supportive of each other. For example, they decided that one rule would be no hitting and punching and they would have to "talk things out" when they became angry. The children also decided that running away from the group was not permissible because it tended to control the group and took time away from planned activities on the field trip.

Every attempt was given to have the child participate fully in the planning process for a field trip. For example, a committee was selected to contact the principal to obtain permission and notify the faculty what they could be planning for the following week. Often the children would not follow through on their assignments, a characteristic of an 8 or 9 year old child. The counselor who was the co-leader in the group would remind the children during the week of their responsibilities and see that they would do the necessary tasks before the next group meeting. Also, the counselor would help the children grasp the implications of committee irresponsibility, should they fail to complete assignments. After 5 meetings, it was obvious the students could assume more responsibility so the counselor diminished her supervisory role.

School personnel frequently question the value of field trips. The following reasons for off campus trips presented to the principal and faculty reduced their skepticism:

1) Some students do not have highly developed verbal skills; hence, field trips offer an experiential learning, situation and concrete basis for verbalization of the experience.

2) The filed trip provides opportunities for more socially adept students to demonstrate skills and serve as role models for less able.

3) The field trip serves as a goal to be anticipated and planned for—and opportunity to work for delayed award.

4) Permission to leave campus provides evidence for Lab students that they are a special group—and a field trip serves to develop group cohesiveness.

5) Field trips are presented to Lab students as evidence that adults have confidence in their ability to represent the school in public.

Field trips were not the only activities with the group. In fact, co-leaders limited field trips to bi-monthly activity. This encouraged the children to think of other activities which could be conducted on the school campus such as kickball, dodge ball, bean bag, and a variety of group games. Children also enjoyed using parts of the session for organized recreation, an excellent vehicle for developing awareness of interpersonal skills. After the on campus recreation activity, the group would "debrief" for about ten minutes to help determine the plans for the following week and to deal with any issues arising during the session, another excellent opportunity to reinforce positive behavior.

Meetings were structured to give students responsibility for planning and initiating suggestions for the group. Each student was elected as "leader of the week" and had the responsibility for conducting the group session. This included appropriate participation from the group members and the responsibility for insuring that group members used a democratic process. At the end of the session, the leader of the week was given a critique by the group members to help him identify his leadership weaknesses and strengths. This technique helped the child discover this capacity for originality, planning, and interpersonal effectiveness. When the inevitable conflicts and disagreements occurred, the adult leaders suggested methods for dealing with impasses or deadlocks and further provided the student leader with feedback on his own approach in handling stressful situations—another example of active adult participants in the group. The co-leaders played baseball, were elected to the teams, participated in field trips, and offered ideas and suggestions during group discussions. However, they withdrew when it became evident the children had learned to be more open in expressing their opinions and ideas and in providing constructive criticism and positive feedback. They also began to withdraw when the children learned how to plan events and assume responsibility.

OUTCOME

The program had impact on the children who participated and in developing a strong collaborative relationship with the school and mental health agency. Teachers and administrators reported a variety of changes in the child's classroom behavior after participation in the group counseling program. Teachers generally agreed that the group participants become more self confident, assertive, and socially appropriate. Questions from children such as "can I be in the 'Lab' next year", "can my friend be in the 'Lab' ", and "why didn't I make the Lab this year" barraged the counselor's office. The program demonstrated to faculty and administration that meeting the affective needs of the child is a responsibility of the school and some children need specialized assistance in this area. The counselor who was a co-leader in the group continued her training in group process and therapy and is presently conducting groups with psychology residents at her school.

This program succeeded in making several of the faculty aware of strengths in a child which could not emerge in a classroom setting. With this awareness, the teachers changed their teaching approaches to focus on where the child was in his development which facilitated his adjustment to the schoolhouse. Most important, the project was a visible concrete example of how a mental health specialist could assist educators by demonstrating his skills in group process and group therapy.

The school district continued its full support of group counseling programs as a result of the pilot project. The groups continue to be conducted on the school campus and they have served as a model for adult team work in developing innovative educational and counseling programs for children. The group co-leaders encouraged parent, faculty, and administrator interaction in order to fully plan and carry out this program. Furthermore, the debriefings, collaboration, and mutual trust and respect which the co-leaders developed for each other demonstrated clearly that working intensely with each other was fun and beneficial for the child.

Authors' addresses: Albert E. Riester, Assistant Professor; and coordinator of Consultation, Community Guidance Center of Bexar County; The University of Texas Medical School at San Antonio, Department of Psychiatry, San Antonio, Texas 78229. Dinah Lee Tanner, Elementary Counselor and, Psychological Associate, Alamo Heights Independent School District, San Antonio, Texas.

Observations on private practice and community clinic adolescent psychotherapy groups

LEWIS H. RICHMOND

Clinical Professor of Psychiatry, University of Texas Medical School at San Antonio: Coordinator of Adolescent Services, Community Guidance Center of Bexar County; Private Practice, San Antonio, Texas.

This paper compares private practice and community clinic adolescent psychotherapy groups. Similarity of composition and difference in outcome of the two types of groups are discussed, as well as unconventional aspects common to both.

THIS paper deals with two groups of adolescents seen by the author, one in a community clinic setting, and one in private practice.

Both groups are open-ended, in which new patients begin, terminate, and drop out at various times. Open-ended groups are preferred because of the heterogeneity of group composition which allows for much sharing of experiences and feelings. This also diminishes resistance for the new member coming into a situation in which he can see peers already established in various stages of trusting relationships with other peers and with group leaders. Co-leaders are utilized so that the adolescent may have access to adult male-female interaction, hopefully on an emotionally adequate basis, with which they can identify.

Each patient is selected for the group after an individual screening. Selection criteria for adolescents admitted to both groups are primarily based on the following: difficulties in expressing thoughts and feelings to parents, problems related to parents who are relatively unavailable physically or emotionally, and problems related to difficulties in separating from families and becoming more independent individuals.

For both groups, the rules are: (1) Whatever goes on in the group stays in the group, unless it is judged to be serious enough (life endangering) that it need be shared with someone outside the group. In such a case, the group member would be told the importance of this, and encouraged to make it known to the appropriate person. If necessary, the leader would assist in doing this. (2) Group members are expected to attend sessions each week and if unable to attend a particular meeting, are to call beforehand so that the group would know why they were absent. (3) If a member does not wish to continue in the group, he is expected to make these desires known to the group prior to his leaving. (4) No drugs are allowed on the premises. (5) Members are allowed to say anything they wish, however they would like to say it; but are not allowed to do whatever they would like. (6) In the past, socializing outside the group was discouraged. However, since it was found that many of the adolescents had known each other prior to being in a group, and others seemed to benefit from social contacts outside the group, this rule was discontinued. (For example, two socially withdrawn boys were able to go to a discotheque together, whereas neither were able to muster up the emotional strength to go alone.) We do ask that activities out of the group be shared during group time so that secrets involving group members be minimized as much as possible. (7) No parents are allowed in the waiting room so that members can feel freer to talk openly.

Of the two groups to be considered, one meets for an hour and a quarter weekly at the Community Guidance Center of Bexar County (CGC) and the other meets for an hour and a half session each week in my office. The teenagers are primarily from upper-lower and middle class families with more lower group income patients in the clinic population. However, class differences between the groups are minimized by a moderate number of military dependents who are seen in the private group. (CHAMPUS Insurance makes private psychiatric help available to lower income families.)

The group at the CGC began on February 1, 1967, and is still in existence. I have chosen to present some observations of the first five years (253 sessions) from February 1967 to February 1972.

The office group began on September 17, 1970 and is still in existence. I will limit my observations to its first two years (102 sessions).

The CGC group began as an all boys group for its first 88 sessions and then became a mixed group after 20 months of operation and has remained so since. The first 22 patients were boys and because of the preponderance of members with inadequate social experiences, it was decided to add girls to the group. During the five years, there were 91 members, 51 of whom were boys and 40 of whom were girls.

The office group began and continues with both males and females, having had a total of 52 members, 25 boys and 27 girls.

The age range was the same for both groups: predominately 15 to 17 year olds with an occasional precocious 14 year old or socially constricted 18 year old.

The majority of patients of both groups were diagnosed as having character and behavior disorders with more boys than girls being withdrawn or schizoid and with more girls than boys acting out sexually or as recurrent runaways. Approximately 70 to 80 per cent of the members of each group had used drugs, many only transiently. Several in each group were referred for depression including suicidal gestures. There were several neurotic diagnoses and an occasional psychotic diagnosis given to members of each group.

In the clinic group there have been over 15 co-therapists during the first five years, ranging in stay from one week to the entire lifetime of the group. These have included staff psychiatrists, psychologists, social workers, psychiatry residents, psychology interns, social work students and medical students. With two exceptions, there was no apparent change in group process with changes in therapists—this may in large part be due to the group's expectation that therapist changes would occur periodically and that at least one therapist would always be constant. The two exceptions were: a psychiatric resident who gave strongly parental responses and who after experiencing the group's anger chose not to return; a medical student who did not respond to specific questions about himself and was chastised for his lack of openness.

The private practice group has had two therapists, one male, one female.

Both groups were led primarily along traditional psychotherapeutic approaches (insight-oriented), but frequently other techniques were utilized when thought to be appropriate. These latter included Gestalt experiences, family sculpting, role reversals (between member and co-leader or between members), confrontation of positive or negative personality features, etc.

We have allowed members to bring a friend or sibling to one meeting provided the guest is acceptable to the group and adheres to the rules of the group. (This practice was begun empirically when a member brought a friend to a group meeting and significant information about the member was revealed by the friend). Fifteen members in the clinic group brought guests, five of whom brought different guests on at least two occasions. Eleven of the 15 members who brought guests were girls and four were boys. In the private group, seven members brought guests—five girls and two boys. The guests were primarily the boyfriend or the girlfriend of the member or a best friend of the same sex as the member. More girls had boyfriends than boys had girlfriends and this apparently explained the preponderance of girl members bringing guests. On rare occasions, guests seemed to be present because of curiosity, but more often because they played an important role in the member's life style. Guests would frequently reveal personal information about the member in a very open way and were rarely present as a "public relations man" for the member.

On at least three occasions, members brought pets (dogs and cats) and this served catalytically to demonstrate which members could readily express warmth or hostility toward the pets.

In the CGC group, 37 of the 91 patients attended over ten group sessions. Of these we consider 20 to have terminated via graduation (making successful progress to warrant no continuation of therapy). Of the 25 who attended between five and ten sessions, nine were considered graduates and 15 dropouts. Of the 29 who attended less than five sessions, 24 were considered terminated by dropping out of the group. (Table I)

In the private practice group, 16 patients attended four or less sessions and four were considered graduates. Eleven attended five to ten sessions of whom six graduated; 25 attended over ten sessions of whom 12 graduated and seven are still in therapy. (Table II)

Although in both groups there seems to be a correlation of successful result with increased number of sessions attended, there was a higher graduation rate for members of the private practice group. Among possible explanations for this are the fact that in the office group: environmental conditions are more private and comfortable (snacks were available—kool aid, cookies); there was less transiency (less total members and thus less dropping out after brief attendance); therapist motivation may be higher because of increased financial remuneration.

In the clinic group, there were less excused absences because often the patients would not call the clinic to indicate when they would miss sessions. In the private practice group, significantly more patients would call if they were unable to come.

In the CGC group there was increased parental pressure for the member to continue in the group as the parents more often felt unable to cope with the child's behavior. In the private practice group there was more parental pressure for the member to drop out of the group because of expenses or because of the family's equilibrium being shifted with the member becoming healthier. Interestingly, several of these members stated that their parents indicated that they were paying two to five times the actual fees for therapy.

All of the above findings are experiential and preliminary observations and it is hoped that a more objective validation will be presented in the future.

TABLE I
Community guidance center group

	Number of members	% of members	Number graduated	% graduated
10+ sessions	37	41	20	54
5–10 sessions	25	27	9	36
0–5 sessions	29	32	5	17
Total	91	100	34	37

TABLE II
Private practice group

	Number of members	% of members	Number graduated	% graduated
10+ sessions	18	41	12	67
5–10 sessions	11	23	6	55
0–5 sessions	16	36	4	25
Total	45[a]	100	22	49

[a] The other seven patients were still in therapy.

Acknowledgement

The assistance of Jan Heyer, M.S.W. and Mary Rice, M.S. is gratefully acknowledged.

Author's address: Oakdell Medical Center, D101; 7342 Oak Manor Drive, San Antonio, Texas 78229

What's in a name?

Name assignment as a pathological function of role confusion in a family

ROBERT L. BECK, PAT WIGGINS and IRVIN A. KRAFT

> The authors examine a process of name and nickname assignment in a family. Particular attention is paid to the way in which the phenomenon in this particular family is a reflection of a firmly established pattern in the father's extended family. Furthermore, there is a review of the beginning stages of their dealing with this issue in family sessions. The paper places emphasis on the accumulation of data related to the naming patterns. In addition, the authors focus on the ways in which the therapists begin to help the child extricate himself from this complex family interaction as a prerequisite for his gaining a greater sense of identity and self-esteem.

A FAMILY often loads the name given to a newborn baby with conscious and unconscious meaning. We suggest the possible symbolism of names and the implicit expectations it carries to the child should not be overlooked in working with a family. These messages—for success, failure, being "like" someone, or the ascribing of some attribute or wished-for attribute—reveal significant undercurrents in the parents' name assignments to new family members.

Berne (1972) examined briefly this issue and discussed it in terms of "scripts," i.e. the living out of family-defined life styles and agendas for living. He stated that:

There is no doubt that in many cases, given names, short names, and nicknames, or whatever praenomen is bestowed or inflicted on the innocent victim, is a clear indication of where his parents want him to go, and he will have to strive against such influences which will be continued in other forms as well, if he is to break away from the obvious hint.

This view of name assignment has also been explored by Seeman (1972):

> Children are named by parents who cherish a set of unconscious expectations for the newborn. These expectations find their symbolic expression in the form of the given name. The final choice which may be a compromise between the current vogue, family tradition, and the unconscious ideas and identifications of parents, brands the child at birth with a tag which symbolizes what his culture and his family expect of him.

We assume here that parents deliberately choose a meaningful name, and, in the following, we explore one case in which name assignment proved significant in the family's way of dealing with one another.

We examine the assignments of nicknames in this family. In its beginning stages, the family's therapy addressed itself to this issue. In our study, one unwilling recipient repeated the pattern with his own stepchild.

We treated this family as part of the overall therapeutic program in our psychoeducational facility. The B.'s, like other families in our program, found themselves dealing with a variety of staff people. In addition to helping the families understand the interrelating nature of our staff members and of our therapeutic interventions, we helped them differentiate the various roles of our staff. This staff network included a child psychiatrist, clinical psychologist, psychiatric social worker, special educators, discovery therapist, art therapist, and a parapsychiatric therapist. Thus, both the child and the family were exposed to a tremendous mix of disciplines; with this mix came a built-in potential for confusion. We found with the B. family an ambiguity and confusion as to who a person was in relation to someone else. They asserted this as a characteristic of their interrelating. Our initial goal, then, aimed to clarify what the family members wanted to be called. We anticipated a considerable amount of confusion on the part of the family members as to who they were and what their roles were in the family, if they themselves remained unclear as to what name or nickname they called each other.

With many of our families, we attempt to extricate the boxed-in family member from the pathological interaction that takes place within the system. Typically, the family member most caught up in this kind of bind has been our identified patient, the child. It seems obvious that until the child knows who he is and makes some contract with himself to take a clear stand on what he is to be called, he will have much difficulty "breaking away," as Berne terms it, from the script his family defined for him. (1972) Stan, who is the center of focus in this family, looked very much in need of differentiating who he was before he could make appropriate moves toward adequate functioning in the outside world, let alone in public school. We set as our goal to break through the confusion of their multiple names and implicit expectations to help Stan get more of a sense of his separateness. We were particularly struck by the fascinating repetition of this process from his stepfather to himself. Our findings illustrate the way in which name assignment and confusion in this family led to identity confusion. Stan, by his symptoms, asked: "Who am I?" "What do I want for myself?" "What script do I follow?"

Our initial diagnostic impressions reflected two concerns: (1) that he seemed to be "in limbo" in school, in peer relationships, and at home; and (2) that he manifested a very poor self concept. The initial consultation indicated that his level of social adaptation was age inadequate characterized by withdrawal, low achievement, and lowered self-esteem. Stan appeared immature and lacking in ego strength. He felt deprived of love, and he seemed inadequate in being able to please anyone.

Links between self-image and name underwent exploration by Bagley and Evan-Wong (1970):

> . . . the majority of previous studies in this area have had to contend with that possibility that the connection between an odd first name and disturbed behaviour could be accounted for by the fact that behavioural oddity in a parent is responsible for both the assignation of a peculiar name and for the child's disturbance.

The authors also view an unusual surname as a possible factor in developing a poor self-image. In addition, they summarized a study done by Boshier:

> Some support for this view has come from a study by Boshier (1968) with New Zealand children, which found that self concept, as indicated by Cooper Smith scale, had a significant correlation with disliking one's *first* name.

Boshier (1968) in this study of eighty children with a mean age of twelve years, used the Cooper Smith scale in rating self-esteem in terms of first names of children. He suggests that "as a relationship between liking one's name and liking or disliking one's self appears, the results seem to support the proposition, at least for the age group studied." In our family, we found such a phenomenon related to the multiplicity of nicknames for this twelve year old boy.

CASE HISTORY

The B.'s came to our program by virtue of the presenting problems Stan exhibited in his school. His principal and teachers both indicated that, while he seemed to be a very bright child, he was sullen, uncooperative, passive, and unwilling to commit himself to doing his work. He usually sat at his desk reading or drawing quietly. He demonstrated an unwillingness to work, to take responsibility for himself, or to respond in any positive way to the educational program. Our staff saw Stan and his entire family, and we recommended that he enter our psychoeducational Day Hospital with his family engaged in our family therapy program. Once with us, Stan exhibited much of the same school behavior. Nonetheless, our educational model, a behavior modification structure, began to meet his passivity head on. Stan incorporated the structure as his own and responded both educationally and behaviorally to the program's expectations.

The B. family consisted of Selma (37), Roddy (43), Stan (12), and Patti (7). Mother and father participated in couple's therapy and Stan underwent individual psychotherapy for approximately two months after entering our program. Prior to the beginning of the family sessions, the family also experienced a two-day Multiple Impact Therapy (MIT) session as modified and practiced in our facility. (Beck, 1972). We decided then not to engage Patti in therapy at this early stage.

Prior to the first family session, the therapists identified a number of concerns related to the interaction of parents and child. Stan resulted from a brief affair his mother had after divorcing her first husband. Her present husband, Roddy, legally adopted him. During their therapy experiences, Stan's parents showed that Roddy was ambivalent about his stepchild's origins and their relationship. As an outgrowth of the MIT, the therapists became confused about what to call Stan, for the family had tossed around a number of names. Somewhere after he came into the school program, Stan became known as Brett; both the teachers and patients called him Brett. At that time the origin of the nickname remained unclear, and Brett-Stan continued unexplored in the MIT and other therapy prior to this first family session.

Family session No. 1

The therapists addressed themselves to Stan and asked him, "What name would you like to go by?" He stated simply that he would rather be called Stan. At this point, Roddy related some history of the name confusion. Stan was the child's given name. His mother's maiden name was Brett; Selma had used it as a nickname at his various schools. She explained that since there were often other Stans in the class, he would, and not often by choice, ask to be called Brett, to be differentiated from the others. Stan let us know that in his new class there was another child named Stan who was quite powerful and who let him know that there would be room for only one child by that name. Nevertheless, Stan stated that he allowed people to call him Brett only when he felt pressured to do so. The name Brett, it should be noted, held considerable significance for Selma, as her maiden name, and as a nickname she felt comfortable in using with her son. Neverthless, Selma mentioned that she never liked it when her *husband* called Stan by this nickname, since he used it as a hostile barb, as a way of putting down *her* family and origins.

Making the issue even more complex, Roddy informed us that he called Stan by another nickname, "Jim." This was sometimes used in another form, "Jimbo." When questioned, Roddy indicated that Jim was the name of an old school and wartime buddy whom he greatly respected and admired. He described his friend as "athletic, daring, and a soldier of fortune," none of which applied to Stan in his stepfather's eyes. Stan affirmed that he did not wish to be called

Jim, Roddy asked him if he *really* had an objection to it, and Stan reaffirmed his position. To this, Roddy asked if "Jimbo" bothered him equally, and Stan replied, "Yes." Two more of Roddy's nicknames for Stan were then uncovered: Bub and Bubba.

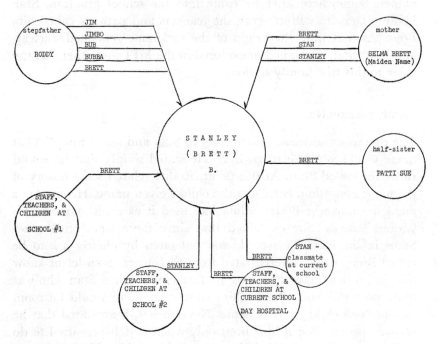

FIGURE 1   Stanley's assigned names at home and school

During the course of our first session, Roddy asked Stan again if the nickname Jim bothered him that much. He talked on about how difficult it would be for him to give up using these nicknames; he repeated again that he never knew that this had been a problem. This session ended with our suggesting we pick up on this again to explore more of the meaning that it had for the family.

Roddy's refusal to call the boy "Stan" suggested his ambivalence about accepting this child and the circumstances of his origins. Stan failed his stepfather's expectations; he could not fulfill the requirements that Roddy maintained for people whom he held in high esteem. The way in which Stan's adoptive father used these assigned nicknames effectively maintained distance and hostile interactions between him and his son. For Stan, the nicknames reinforced his stepfather's unyielding expectations and refusal to accept Stan as he was. In addition, the assignment of some nicknames, particularly Brett, constantly reminded Stan to whom he really belonged. He realized the long way that he would have to go until he would meet his father's expectations.

Family session No. 2

In this session with the family we studied the history of nickname assignment in their extended families. In Selma's family, nicknaming virtually did not exist. In Roddy's family, it was rampant and extensive. The father's first name, Roddy, duplicated that of a wrestling duo, as his father was very fond of the sport. His middle

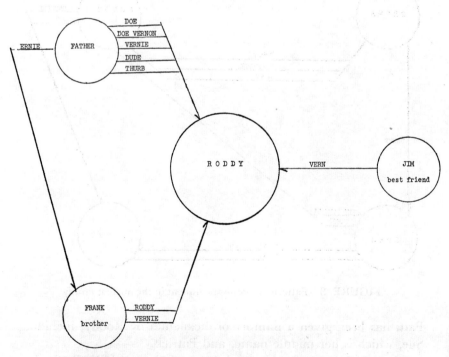

FIGURE 2 Step-father's assigned names in his family of origin

name, Vernon (Vern), was often used as a nickname by his father and his best friend, Jim. His father also called him "Doe," "Dude," and "Thurb." While Roddy feels that none of these names meant derogation, he clung to the name Roddy in school. "I did not want to be called a nickname by anyone other than close friends or family." His brother Frank received a variety of nicknames also. One of them included the name of another wrestler. Roddy also stated that his father lived with numerous nicknames.

Roddy calls his wife by a number of nicknames: Sel, Sweetie, Baby, and Buger. In turn, Selma calls him Honey and Roderick (usually in anger).

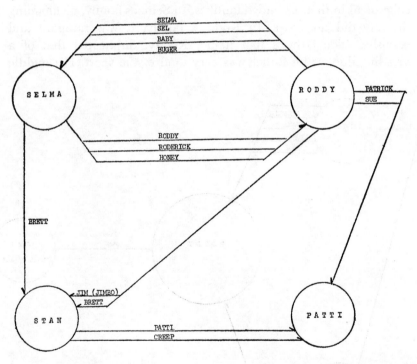

FIGURE 3 Patterns of name assignment in the nuclear family

Patti has been given a number of nicknames by Roddy, including Sue, which is her middle name, and Patrick.

Stan's nicknames for his mother and father and Patti's for her parents seem rather traditional, with "mom" and "mother" and "dad" and "father" used interchangeably. Stan calls his sister "creep" and "Patti." "Creep" is used when he is angry and putting her down. (see Figure 3).

We found particular interest in the pattern of nickname assignment beginning at least one generation back with Roddy's father. Roddy expressed some feelings about being called these nicknames as it related to his own situation with his father. When he told the story of his being named for a wrestler, he noted that while his brother had also been so named, the tag was only a nickname, and not a given name. "Somehow," he said, he was "stuck" with the name.

CONCLUSION

We have chronicled these interesting and complex transactions evolving out of unwanted names in the B. family. We uncovered this phenomenon in the B.'s when we found ourselves struck by our own uncertainty as to what to call Stan. While we did not elaborate on the post-negotiation repercussions of the two sessions with the family, we can mention that Stan ultimately communicated his own wants first to one member of the educational staff, and then to all members and the other children in the day program. As of this writing Stan is called Stan by classmates and teachers. He is making movement both in family therapy and in his educational situation.

We have focused purely on one family. We also noted the ways in which clarification of what this child was to be called affected how he saw himself. The family's negotiating this formerly "taboo" issue seemed then to have an effect on Stan's own desire to assert himself more, to move out of this passive and oppositional stance, and to make some movement toward achieving and utilizing his abilities.

Of particular importance is the significance names have in this family, both in terms of the generational repetition and in terms of the vise-like grip the ambiguity seems to have had on this child. We suggest that the material presented indicates that some consideration should be made in working with families as to family names and nicknames as part of their initial diagnostic process. In a setting like ours, with so many disciplines and so many people involved, it is tremendously important to get clarification on this kind of issue prior to having any expectation that there be therapeutic or educational movement for any family member.

Acknowledgement

We wish to express our appreciation to Jay C. Maxwell, M.D., for his ideas and support of this study.

References

Bagley, C. and Evan-Wong, L. (1970): Psychiatric disorder and adult and peer group rejection of the child's name. *Journal of Child Psychology and Psychiatry*, 11, 19–27.
Beck, R. L., Vick, J. W. and Kraft, I. A. (1972): Pathosyntonic families in a psychoeducational program, unpublished paper presented at the Southwestern Meeting of the American Orthopsychiatric Association, Galveston, Texas, November, 1972.
Berne, E. (1972): *What Do You Do After You Say Hello?*, Grove Press, New York, 78.
Boshier, R. (1968): Self-esteem and first names in children. *Psychological Reports*, 22, 762.
Seeman, M. V. (1972): Psycho-cultural aspects of naming children, *Canadian Psychiatric Association Journal*, 17, (2), 149–51.

Authors' addresses: Robert L. Beck, A.C.S.W., Psychiatric Social Worker, Texas Institute of Child Psychiatry, at Texas Children's Hospital, Houston, Texas 77025. Pat Wiggins, Clinical Psychologist, Texas Children's Hospital, Houston, Texas 77025. Irvin A. Kraft, M.D., Medical Director, Texas Institute of Child Psychiatry, at Texas Children's Hospital, Houston, Texas 77025.

Homosexuality: a confused trinity

SIDNEY J. FIELDS

Homosexuality is discussed as a generic term which describes three separate entities rather than a single process. Research and clinical evidence is cited to support this position and the usefulness of making this distinction is discussed.

ALTHOUGH homosexuality is commonly talked about and read about by both the general public and professionals of all sorts, it remains a much misunderstood topic. And when we hear the term homosexual applied to an individual as though to describe him there is either a strong implication or a strong inference that everything of importance has been said that needs to be said about that person, that the one word conveys all. We are led to believe that we are dealing with a single, all-embracing entity. Indeed, the current Diagnostic and Statistical Manual of Mental Disorders (DSM II) of the American Psychiatric Association insists that Homosexuality be identified as a separate category under Sexual Deviations. It allows no distinctions.

But in clinical practice when we try to keep to that idea of homosexuality as a single entity we frequently find outselves faced with contradictions and inconsistencies. On becoming aware of this we begin to feel puzzled, increasingly frustrated, and generally at a loss. Recourse to the library to review the literature doesn't help much. What research has been done seems to yield ambiguous results, except for one notable exception which we will mention. Our time searching the literature appears largely fruitless.

The confusion begins to clear when we question whether homosexuality is, in fact, a single entity.

We then discover that we are not, apparently, dealing with a unitary condition at all. And we recall the famous first sentence of

Caesar's Commentaries, curiously suggestive here: "Omnia Gallia in tres partes divisa est." That could be paraphrased in the present context to read, "All Homosexuality is divided into three parts." Seeing homosexuality only as a unitary concept leads to misunderstanding and confusion.

The situation is like one familiar in general medicine where the first symptom the physician detects is a fever. This alerts him and indicates something is going on with the patient. The physician doesn't yet know what that something is because an elevated temperature in itself has little meaning. The patient's true condition cannot be understood or diagnosed on the basis of the fever alone. Yet a longstanding premise in medicine states that the presence of a fever is a reflection of existing disease process within the body. Then all effort is turned to the search for the origin of the fever. When that underlying process is located we are at once in a position to understand both the nature of the disorder and the specific significance of the fever.

Countless physiological disorders can give raise to elevated temperatures. The symptom of homosexuality appears, fortunately, to derive from a very limited number of basic personality configurations. We can distinguish three at the present time. These are the invert, to use Havelock Ellis' term; the paranoid personality; and the antisocial personality. The nature of the homosexuality embedded within each of these three personality contexts is designated essential homosexuality, latent homosexuality, and incidental homosexuality, respectively.

The invert is the exclusive homosexual. Ellis chose the term invert deliberately to avoid pervert and the activity as perversion. He did so after Freud defined perversions to be all sexual activities among adults except the union of male and female genitals. Freud (1930) further stated that "neurosis is the negative of a perversion." This means, Clara Thompson (1950) explains, that "neurotic symptoms represent repression of perverse sexual interests. Perversion, on the other hand, does not spring from repression. In the perversions infantile sexual interests remain conscious and receive gratification. Because there is satisfactory discharge of the libido there is no damming of energy and repression does not take place. This logically brought Freud to the conclusion that in the case of perversions there is no neurosis and nothing can be analyzed."

To the extent that the exclusive homosexual, the invert, is not conflicted about his homosexual orientation and is therefore not neurotic about it, to that extent the statement is true and accurate. However, the exclusively homosexual individual may be conflicted and therefore neurotic in areas of his life having little or nothing to do with his homosexuality. Freud clearly concurred with this point of view when he stated in a letter to an American mother regarding her son's homosexuality: "What analysis can do for your son runs in a different line. If he is unhappy, neurotic, torn by conflicts, inhibited in his social life, analysis may bring him harmony, peace of mind, full efficiency, whether he remains homosexual or gets changed." (SIECUS, 1970).

It is for these difficulties in other areas of his existence that the invert personality may seek and benefit from psychotherapy, at least to the extent that heterosexual patients do. He or she is characteristically content with his or her† homosexual orientation. This is not what bothers him and he wants it left alone. Indeed, it may be for the very reason that the invert personality does accept his homosexuality so openly that he is rarely seen in outpatient clinics or as admissions to psychiatric wards. Again, when he does appear in these places it is apt to be for reasons other than his homosexuality.

The findings of Dr. Evelyn Hooker's (1956–1957) landmark study are well worth noting here. She wished to determine whether or not homosexuals are as adjusted in their lives as are heterosexuals in theirs. But she avoided the traditional confusion of randomly lumping all "homosexuals" together by carefully selecting for study only those male homosexuals who openly identified themselves as such, whose patterns of sexual desires and overt behavior were predominantly or exclusively directed toward members of their own sex. who were not seeking psychological help, and who were gainfully employed. It happens that these criteria describe precisely the

---

† For ease of presentation I will use the masculine pronoun (he, him, his) with the understanding that the reference is intended to include the female homosexual as well.

first of the three personality categories we are delineating here, the one referred to as the invert personality whose homosexuality is essential and exclusive. She proceeded to match each of the 30 male overt homosexuals in her sample with 30 heterosexuals on the basis of age, education, and IQ. Then she administered a series of projective techniques and attitude scales and gathered further information from intensive life history interviews. All this material was presented to a panel of experienced clinicians with the request that they rate each subject on a 5-point scale of adjustment. The specific sexual orientation of each of the subjects was, of course, withheld from the judges. The findings were clear and unambiguous: the ratings of adjustment for homosexuals and for heterosexuals were not significantly different. Indeed, the judges found it impossible to distinguish between the two groups. They were generally unable to pick out the homosexual person in the 30 matched pairs.

A brief description of the personality of the typical exclusive or essential homosexual is appropriate at this point. He is apt to be above average in intelligence, with a level of education commensurate with it. He tends to be of a refined nature, preferring the more cultured tastes. Hence he is drawn to the arts, though the social sciences and religion attracts him too. He is capable of strong devotion and loyalty. It appears likely that his homosexual affairs and marriages are as stable and enduring as are hetrosexual affairs and marriages. He can be just as downcast and despondent over the loss of a homosexual lover as the heterosexual can be on losing his lover. Jealousy is as common among essential homosexuals as it is among heterosexuals. He is gregarious, seeking especially the company of other homosexuals, though he also enjoys the company of other persons of culture and refinement who are able to accept him as he is. Should psychopathology develop it is likely to be mild psychoneurosis, not centered primarily around his homosexuality, or a psychosomatic condition. Psychosis occurs rarely.

The homosexuality which appears in the context of the paranoid personality is distinctly different from essential homosexuality as described above. This homosexuality is the latent homosexuality of traditional psychiatry. The individual with latent homosexuality cannot accept even the thought of homosexuality. Certainly any vestige of it in himself or any implication that it might be present becomes a source of constant threat to him.

Classical psychopathology presumes that embedded within every paranoid condition lies a homosexual conflict. Experienced clinicians have encountered enough exceptions to this dictum to question its generality. Though homosexual conflict does occur in many paranoid cases, still there are instances in which no such conflict can be elicited. To expect to find it in all cases is definitely misleading. The old assumption, therefore, looks more and more like a myth. No useful purpose is served in continuing it.

But the paranoid personality which harbors latent homosexuality does live under ceaseless threat. The threat is perceived by the patient as coming from two directions. One is from within the self, at various levels of awareness. The other direction is from outside the self, from the social environment surrounding the individual. When the living situation is such that the paranoid personality with latent homosexuality is forced into close and relatively intimate quarters with other men, or in the case of female homosexuals of this type with other women, the threat intensifies and may become intolerable. The individual passes from mild unease to discomfort to agonized distress, until at last his ego becomes disorganized, he "goes to pieces" and is left in the fearfulness of homosexual panic. The military services are well acquainted with this sequence of events among their personnel. Barracks life with its confinement, where all must share the same bed space, toilet and bathing facilities, is ultimately unendurable for the person with latent homosexuality.

The paranoid individual with latent homosexuality will usually go to great lengths to assure himself and others that he is not "tainted" with homosexuality of any sort or in any degree. Indeed, this effort becomes a central theme in his life. A frequent strategy is to seek marriage with a person of the opposite sex. The chosen partner is likely to take the marriage in good faith, without the faintest idea that she (or he) is serving a special purpose. She may become aware over time that he has little emotional warmth for her, that he seems to prefer distance to closeness. He may want a child or children, but they, too, are for display purposes only. If his wife questions him he is apt to aver that he is content with things as they are, can see no reason for change, and won't speak of divorce. The underlying reason for his attitude, of course, is that the marriage stands as self-evident proof to himself and others that he is fully and entirely heterosexual. The symbolic value of the marriage for him is therefore priceless. It follows that he must keep the marriage intact at all costs. He hangs on to it with desperation. Happiness is not what he looks for in marriage. What he seeks is simply some measure of relief from the relentless threat of his latent homosexuality.

The third basic personality structure associated with homosexual behavior is the classical psychopath, more recently called the sociopath, and currently termed the antisocial personality. The homosexual behavior which occurs in connection with the antisocial personality is almost accidental in the sense that it is determined more by random chance than by deliberate preference. We designate this type of homosexuality incidental. A closer look at the general characteristics that define the psychopathic personality will make this clear.

The psychopath does not feel bound by the conventions of society. He finds himself therefore in the basic position of being free to do very much as he pleases. However, since he is not psychotic and remains in close touch with reality, he does keep a wary eye on the law as he moves along. He is deficient in conscience, so experiences little or no primary or neurotic guilt. He is capable of feeling guilt of a different sort, however, a kind of secondary guilt. It is what he feels when he gets caught for illegal or immoral activity. He berates himself for getting caught, not for acting wrongly. His tolerance for frustration is notoriously low. He cannot endure for long putting off what he wants right away. Whatever it is, he must have it now. His every need must be gratified promptly, whether it be for food, drink, money, sleep, sex, or anything else. When sex is the target usually any compliant partner will do. For this reason he could as accurately be thought of as ambisexual for it is largely a matter of indifference to the psychopathic personality whether that partner happens to be of the opposite sex or of the same sex. Obviously, when the partner happens to be of the same sex the relationship is homosexual. It could as easily be the other way around. This is why the adjective incidental or accidental seems so fitting for this type of homosexuality. But regardless of the sex of the partner of the moment, the relationship is typically short and transitory, a one-night stand. The psychopathic personality shows little or no emotional investment in his partner, male or female. He doesn't want to be tied down in any way or to any one. He is also boastful, inclined to brag freely and openly about any of his activities when he feels it is safe to do so. He relates his sexual exploits with equal glee and gusto, with little consideration for the feelings or welfare of his partner. And since he does not hesitate to take full advantage of his partner these sexual encounters frequently become sadistic and murderous, depending on the impulse of the moment.

The table shown here summarizes the differential features of the nature of the homosexuality associated with each of the three basic personality types. These features stand out most clearly as one reads down the columns. Thus the first column, headed Partner Choice, indicates that the person with Essential Homosexuality will choose a partner of the same sex, while the Latent Homosexual will choose a partner of the opposite sex, and the Incidental Homosexual will accept a partner of either sex.

The second column describes the homosexual's emotional attachment to the partner of his choice. The Essential Homosexual's attachment to his partner is firm and enduring; of the Latent Homosexual to his partner loose, yet enduring; and of the Incidental Homosexual to his partner superficial, temporary, and changing (or short, shallow, and shifting).

The next column headed Ego Reaction deals with the attitudes of the three categories of homosexual toward conscious awareness of their homosexual inclinations. We see that the Ego Reaction of the Essential Homosexual is conscious acceptance, that is to say, ego-syntonic. By contrast, the Latent Homosexual's Ego Reaction is conscious rejection, or ego-alien. And for the Incidental Homosexual his Ego Reaction is conscious acceptance (the antisocial personality doesn't care who knows about his sexual behavior, really, provided the police are not listening).

With knowledge of the particular Ego Reactions one can readily infer whether guilt feeling is present or absent. The next column

TABLE I

Differential homosexual behavior associated with three basic personality structures

Basic personality	Partner choice	HOMOSEXUAL BEHAVIOR Partner attachment	Ego reaction	Guilt feelings	Pathological outcome
INVERT ESSENTIAL HOMO.	Same sex	Firm and Enduring	Conscious acceptance (Ego-syntonic)	Absent	Mildly psychoneurotic (jealousy)
PARANOID LATENT HOMO.	Opposite sex	Loose and Enduring	Conscious rejection (Ego-alien)	Present	Homosexual panic and paranoid psychosis
PSYCHOPATH INCIDENTAL HOMO.	Either sex	Superficial, temporary, changing. (Short, shallow, shifting)	Conscious acceptance	Absent	Antisocial, sociopathic personality

notes the presence or absence of guilt feelings for each of the three homosexualities. The Essential Homosexual experiences little or no guilt concerning his homosexuality. By contrast again, the Latent Homosexual is plagued by a strong, almost inescapable sense of guilt. The Incidental Homosexual, on the other hand, displays no guilt about anything in his antisocial life, including his sexuality, homosexual or otherwise.

What is the outcome likely to be if psychological stress develops and intensifies in each of the three homosexual configurations? The last column suggests what may be expected. The Essential Homosexual may develop a mild neurosis or psychosomatic condition, with jealousy and despondency frequently present. The psychopathological outcome of the Latent Homosexual is homosexual panic and paranoid psychosis. The Incidental Homosexual seems simply to reinforce his basic personality to become even more overtly sociopathic. He'll have a harder time evading the law. And the frequency of unnatural deaths will be greater among sociopathic homosexuals than among the other two categories.

What I have presented here is a descriptive frame of reference within which three kinds of homosexual behavior become understandable when seen in the context of one of the basic personality configurations of which they are a part. Nothing is said concerning origins and etiology; that remains for further inquiry. Yet experience in teaching in undergraduate and graduate medical education shows this descriptive framework to be clinically useful for several reasons. It removes homosexuality as a hazy, shapeless concept and replaces it with three fairly well-delineated patterns of attitudes toward persons of the same sex, with each of the three taking meaning from the basic personality types of which they are part. Our diagnostic conclusions, based on firmer understanding, are sounder. And consequently we are in better position to determine appropriate dispositions. It is as important to realize when to let a thing alone, as the Essential Homosexuality of the invert personality, as to know when to treat it with delicacy and sensitivity, as with the Latent Homosexuality of the paranoid personality. And it is of genuine therapeutic value to the patient to experience the confidence that comes when he feels he is being understood by his psychotherapist.

F

Lastly, we may anticipate that research studies which take these distinctions into account in the selection of experimental populations will be rewarded with higher yields of clean-cut, consistent findings.

References

Ellis, Havelock.: *Studies in the psychology of sex.* Vol. 2. New York: F. A. Davis.
Freud, Sigmund. (1930): *Three contributions to the theory of sex.* New York: Nervous and Mental Disease Publishing Co., 28.
Hooker, Evelyn. (1956): A preliminary analysis of group behavior of homosexuals. *J. Psychol.*, **42**, 217–225.
Hooker, Evelyn. (1957): The adjustment of the male overt homosexual. *J. Proj. Tech.*, **21**, 18–31.
Sex Information and Education Council of the U.S. (SIECUS) Newsletter, (1970): **6**, 5.
Thompson, Clara. (1950): *Psychoanalysis: evolution and development.* New York: Hermitage House, 43.

Author's address: Associate Professor, Department of Psychiatry, University of Arkansas Medical Center, 4301 West Markham, Little Rock, Arkansas 72201.

The concept of identification: an evaluation of its current status and significance for group psychotherapy

MORTON KISSEN

> The present paper involves an attempt to present a conceptualization of the process of identification as it occurs during the course of group psychotherapy. The positions of psychoanalytic theory, learning theory, social learning theory, and the psychology of cognition with regard to identification will be explored and integrated into a unified conceptual framework.
>
> The rather broad conception developed in the present paper will highlight the importance of specific model attributes, particularly as they relate to characteristics of the group therapist that may enhance identificatory interactions. Four testable hypotheses with regard to the physical characteristics, physiognomic characteristics, psychological characteristics, and sociological characteristics of the group therapist will then be presented as potentially important factors in determining whether or not an identification with the therapist by the group members occurs.

THE concept of identification has always been a central one in the study of the psychotherapeutic process. It is widely agreed that, at some point during psychotherapy, the patient takes over for himself a number of the personality characteristics of his therapist. Patients are typically more similar to the therapist at the termination of therapy than they were at the outset. They have been found to internalize personality features such as the expressive and cognitive style, defensive structures, broad character patterns, and even a number of discrete personal traits and behavior patterns of their therapists. There is no reason to assume that such a process of internalization is restricted to individual psychotherapy. A similar process may be inferred as occurring during the group psychotherapy interaction. Individual group members, to varying degrees, presumably internalize personal characteristics of the group therapist.

The nature of the latter process of internalization is only vaguely understood at present. Attempts have been made to conceptualize the process from a variety of vantage points. Thus, psychoanalytic theory (Brody, 1964; Freud, 1960; Freud, 1933; Freud, 1953; Freud, 1953; Knight, 1940; Schafer, 1967; Singer, 1965), learning theory (Dollard, 1941; Lazowick, 1955; Mowrer, 1950), sociological and social learning theory (Bandura, 1963; Bandura, 1963; Bandura, 1961; Parsons, 1955; Sears, 1962; Whiting, 1962), and the psychology of cognition (Kagan, 1958), have all, at one point or another, served as the framework for an attempt at conceptualizing the process of identification. Although a great deal of overlap can be noted between each form of conceptualization, a unifying integration of all of them has not been attempted. In the present paper, the contribution of each of the previous approaches to the study of identification will be highlighted. A conceptual integration will subsequently be offered, as well as a number of hypotheses with regard to group psychotherapy interactions stemming from such an integration. The hypothesis will be focussed at the characteristics of the group therapist that are associated with heightened identificatory behavior.

THE PSYCHOANALYTIC CONCEPTION OF IDENTIFICATION

Freud placed the concept of identification within a *group* context in his volume entitled *Group Psychology and the Analysis of the Ego*. (1960). He referred therein to a process by which the individual members of a group tend to become an integrated unit via a common internalization of the leader. Thus, he defines a group as—". . . a number of individuals who have put one and the same object in the place of their ego ideal and have consequently identified themselves with one another in their ego" (Freud, p. 61, 1960). The members of a group, according to Freud, establish a regressive libidinal attachment to the leader, much as they had done to their parents at an earlier age.

Under the spell of a strong leader, the group member becomes a dependent child.

While the members are bound by libidinous ties to their leader, the leader's psychology sharply differs from that of the members. He has no emotional attachments to anybody but himself and it is precisely this *narcissistic* quality which makes him a leader. Freud does not envision a leader whose political psychology is democratic in nature. He rather pictures a forceful, masterful, brutal, and highly narcissistic individual who uses his power to bend the members of the group to his will.

Identification, for Freud, at least within a group context, involves a regressive process of interaction that occurs when a group of individuals meet together with a strong and autocratic leader. The latter process of regression is coupled together with a loss of intellectual capacity, a heightened suggestibility, and a sort of overall state of *fascination*, not unlike the trance-like state induced by an effective hypnotist.

Freud elaborates upon essentially two different conceptions of identification in a series of writings (1960, 1933, 1953, 1953). In one conception, he speaks of identification as a form of restitution of lost love objects. Thus, the lost object is assumed to be introjected into the ego with the ego assuming the characteristics of that object. The latter introjected characteristics are further assumed to become the nucleus of the emerging superego. There is a regressive character to this conception of identification, the latter tending to resemble a primitive oral introjective process. Freud distinguishes between the more developmentally sophisticated *object* choice in which the child wishes to *possess* his parent and the more primitive state of identification in which the child wishes to *be like* his parent. Nevertheless, the latter involves a subtle process of ego alteration, in which the child takes over characteristics from his parents. Thus Freud, in speaking of character development, states—"The incorporation of the early parental function in the shape of the super-ego is no doubt the most important and decisive element; next come identifications with the parents of a later date and with other parents in authority, and the same identifications as precipitates of abandoned object-relations." (Freud, p. 126, 1933).

Freud propounds still another conception of identification. The latter is contained in the notion of *identification with the aggressor*, developed to explain the method by which the child resolves the oedipal crisis. In the narcissistic interest of protecting his penis, the child gives up his attachment to his mother, replacing the latter attachment by an identification with his father. Freud writes: "The authority of the father or the parents is introjected into the ego and there forms the kernel of the super-ego which takes its severity from the father, perpetuates his prohibition against incest, and so insures the ego against a recurrence of the libidinal object-cathexis." (Freud, p. 273, 1953). Thus, in males a civilizing process occurs as a result of the castration threat and consequent identification with the threatening father figure. Ego and superego development, for Freud, hence consists of a departure from primary narcissim brought about by libido being displaced to an externally-imposed ego-ideal. The renouncing of instinctual impulses proceeds precisely through an identification with the culture and its most potent transmittors. Singer, in a lucid but not altogether convincing article (1965), argues that such a notion of identification implies a form of externally-imposed social conformity and loss of identity—as opposed to a more spontaneous process of self-determination.

Brody and Mahoney, in a very important article (1964), make the critical distinction between processes of identification at early and late stages of ego development. They state that prior to the formation of the mature ego, perceptions are diffuse and there is no awareness of a self-object differentiation. Perceptions are, at that time, assimilated through primitive processes of *introjection*. Later on in development, when the ego is more sophisticated and capable of realizing that the object is external to itself, the process is referred to as *identification*. Thus, Brody and Mahoney place the concept of identification within a *developmental* context. They distinguish between a process of assimilation occurring at a developmental stage during which the ego is weak and only partially differentiated from the object world (introjection) and a more differentiated form of assimilation occurring at a later stage in which the ego-object differentiation has been firmly established (identification).

They make still another important distinction between identification as a *process* of ego development and as an *end result* of such development. Thus, they refer to identification, in one context, as the assimilative "process" by which the ego is gradually formed during the course of development. The process of additive ego alteration is caused, according to them by ". . . an integration of highly complex mental functioning, including memory, retention, conceptualization, degree of consciousness, and perception." (Brody, p. 61). In still another context, they refer to identification as the summated "content" of the afore-mentioned developmental process. Identification, in such a framework, involves the ego and superego as formed and organized structures, as well as their internalized contents (values, expressive styles, impulse-defense configurations, etc).

Schafer (1967) enlarges the psychoanalytic conception of identification beyond the depiction of a process in which the child internalizes parental superego characteristics (instinctual renunciations, moral judgments) to one in which the child internalizes an integrated composite including ego and id attitudes as well. He states: "As in the case of ideals, this separation of identification is essentially a conceptual artifact. For what do we observe in our patients? We find that what they have identified with are complex id-ego-superego positions or mental organizations—particularly those of their parents. For example, in the identification with a parent given to outbursts of temper, it is not only discharge of id aggression that is involved; also included in the identification are ego positions on defense and control and superego positions on prohibition and renunciation" (Schafer, p. 142). Thus, in Schafer's conception, total structural constellations of the parental figures come to be duplicated in the child.

Knight (1940) adds still another dimension to the conception of identification, speaking of an interaction of introjective and projective processes. Thus, he argues that what exists in reality are complex introjective-projective processes, with one or the other predominating at a given moment of time. Knight thus thinks of identification as ". . . the result of various mechanisms, introjection being the principal one, but with projection, displacement, substitution and perhaps other mechanisms also in operation." (Knight, p. 336).

Psychoanalytic thinking has tended to be most incisive when describing the *end products* of identificatory processes such as ego ideals, impulse-defense configurations, and superego constellations. There has been slightly more fuzziness, however, in the analytic description of the *process* by which such structural contents come about. Psychoanalysts have largely restricted themselves to vague metaphorical analogies to oral incorporative processes such as biting and swallowing. The motivation for identification is, at times, assumed to be the threat posed by a potentially aggressive adult authority and, at other times, assumed to be an apprehension over the possible loss of love of a warm and supportive love object. Very little clarification of the exact mechanics of the process by which such motives lead to established identifications is offered, however.

LEARNING THEORY FORMULATION

The learning theory approach to the problem of identification has tended to focus upon an explication of the mechanism by which identification takes place.

Dollard and Miller (1941) propound a theory of imitation learning that, at least outwardly, appears relevant to the concept of identification. According to their theory, imitation learning consists of the positive reinforcement of a motivated subject for matching the correct responses of a model during a series of initially random, trial-and-error responses.

Although Dollard and Miller consider their illustrative experiments as demonstrations of imitation learning, Bandura (1963) has argued that they in fact represent only the special case of discrimination place learning, in which the behavior of others provides discriminative stimuli for responses that already exist in the subject's behavioral repertory. Such a theory, according to Bandura, cannot account for the occurence of imitative behavior in which the observer does not perform the models' responses during the acquisition process and for which reinforcers are not delivered either to the models or the observers. At any rate, Dollard and Miller's theory must be viewed as relevant only to circumstances in which the observer is blind to the full nature of the stimulus circumstances and hence must depend upon a *matching* of the model's behavior in order to attain a rewarding outcome.

Mowrer's theory of imitative learning (1950) emphasizes the importance of the principle of *contiguity*. He describes a form of "vicarious learning" in which the response-correlated stimuli of the model arouses in the observer an expectation that he too will experience analogous response-correlated stimuli for acting in a manner similar to the model. Thus, when the observer sees a film model's verbal and behavioral expressions of satisfaction after emitting a particular response he is quite likely, according to Mowrer, to imitate that response. Mowrer argues that the learned imitative responses are sustained in the observer by a rewarding (motivational or emotional) proprioceptive feedback that has been associated through classical conditioning with that particular emitted response. Thus, imitation occurs only when the observer is directly or vicariously rewarded by the sensory consequences to himself of the model's instrumental responses.

Mowrer, in addition, describes a form of imitation learning in which the model directly rewards the observer while, at the same time, emitting a particular response. The model's response hence tends to take on a *secondary reward* value for the observer leading to attempts by the observer to approximate that response at times when it is not being made by the model.

The learning theory approach to the problem of identification has frequently been criticized for its *hyper-specificity* and elementaristic slant. Thus, there appears to be an emphasis upon the learning of specific S-R connections rather than upon the incorporation of total and complexly-organized behavior constellations. Lazowick (1955) describes a learning conception in which a subject matches the *mediating responses* or complexly-integrated "meaning systems" of the model rather than merely matching the discrete responses emitted by the model. In contrast to a mere *imitation* conception in which tiny and discrete behavioral responses are incorporated into the observer's response repertory, Lazowick proposes a concept emphasizing the incorporation of broad cognitive meaning systems.

Such a conception, he feels, is much more relevant to the broad characterological, expressive, and moral structural changes that lie at the heart of the concept of identification.

SOCIOLOGICAL AND SOCIAL LEARNING THEORIES

The sociological conception of the identification process emphasizes that identifications occur within a social interaction matrix. Family interactions are seen as a crucial determinant of internalization processes. Thus, the total family is considered to be a complex interactional framework within which the child learns and hence takes over a variety of role identifications.

Sears (1962), studying the impact of particular child-rearing practices upon conscience development in children, concludes that the development of an adequate conscience requires an initial stage of dependence upon parents who are warm and affectionate within a family atmosphere of mutual trust and esteem in which love-oriented techniques of discipline are used. The child who is brought up in such a home adopts the parents' values and ideals as a part of his own charter of conduct in order to insure a continuation of his parents' love.

In the parent-child diad or interaction, the parent tends to have direct control over many *resources* which the child is capable of controlling only indirectly through the parent. The capacity of parents to mediate resources is an essential factor in Whiting's (1962) theory of identification. Whiting speaks of an initial *cognizance* stage in which the child develops the capacity to predict reciprocal adult behavior accurately. The child tends to envy the status of his parents whom he perceives as having more control over resources than he has. As soon as envy occurs of a given status, the child, according to Whiting, attempts to play the role associated with such a status. The child covertly practices the role, fantasizing himself as the envied person. He also practices the role overtly. The child during the series of processes depicted by Whiting, gradually takes over the disciplinary and moral roles of his parents.

Perhaps the most elaborate and empirically-validated conception of identification is contained in Bandura's writings (1963, 1963, 1961). Central to Bandura's approach is his emphasis upon the *modelling* process. Children, he feels, are capable of a form of observational learning in which they directly take over a number of traits, attitudes, and behavior patterns from adult models, without the mediation of reinforcements. Thus, neither the child nor the adult model need be rewarded for social learning to occur. The mere observation of an adult model's behavior sequence can, under certain circumstances, lead to an enhancement of that form of behavior in the child observer. To substantiate the esixtence of forms of social learning unmediated by reinforcement contingencies, Bandura presents a number of experimental demonstrations. In the demonstrations, children are shown to take over a variety of behavior patterns such as aggressive activities and patterns of self restraint.

Bandura does not, of course, deny the importance of reinforcement contingencies upon the modelling process. He feels, however, that rewards and punishments affect the *performance* of a given behavior pattern more than they do the initial *acquisition* of that pattern. Imitative responses are acquired, according to Bandura, primarily as a result of the contiguity of sensory events. Quite complex and totally novel behavioral sequences can be reproduced by children after merely a single observation of an adult model in action.

Bandura gives a number of examples of situations in which imitative social learning can run counter to standard social training practices. Thus, he describes the case of a parent who physically punishes a child for having struck a neighbor's child, and states: "Concurrently with the intentional training, however, the parent is providing a model of the very behavior he is attempting to inhibit in the child." (Bandura, page 1, 1963).

The primary contribution of Bandura's approach is his strong emphasis upon a form of social learning, directly *cognitive* in nature, and yet responsive to the standard learning principles of reinforcement. Identification, according to his conception, is thus a form of social learning in which the child imitates novel patterns of behavior that he has perceived in the adult models of his environment.

A COGNITIVE APPROACH

Kagan (1958) thinks of the identification process as involving an empathic cognitive bond between subject and model. The subject is motivated toward experiencing or obtaining positive goal states which he perceives that the model commands such as mastery of the environment and love and affection.

At some point during development, the subject is reinforced by the environment, according to Kagan, to believe that being similar to an esteemed model is equivalent to possessing his positive and desirable characteristics. Thus, the identification response is secondarily reinforced by perceptions of similarity between characteristics of the subject and model. An identification is maintained only so long as the subject perceives the model commanding desired goal states. The subject is hence constantly directed toward bringing his own self perceptions into congruence with those of an esteemed model with whom an empathic bond has been established. He attempts to establish behavioral similarities to the model, in some cases actively imitating activity patterns.

Kagan feels that identification tendencies should decrease with age as the subject begins to no longer need to gratify mastery and love needs through a "vicarious" mechanism. Behavioral tendencies that have been established as a result of identification can then function autonomously within the personality of the subject.

A CONCEPTUAL INTEGRATION

The theories of identification discussed above are all attempts at explaining the phenomenon by which one individual takes over attributes from another during the course of an interpersonal relationship. It is highly unlikely that any one of them is sufficient to deal with the phenomenon in and of itself. It is much more likely that an integration of all of them will be necessary, with an emphasis being placed upon both *process* and *content* aspects.

Identification, when viewed in terms of *process*, has structural, defensive, and cognitive characteristics, and is intrinsically related to social learning and character development.

The *structural* aspect of identification is clearly implied by its intimate involvement in the process of ego-formation. The very nature of the ego is shaped through identifications. The introjection of characteristics possessed by parental and other forms of authority figures helps to shore up the ego and give it substance. At moments of great object loss, introjection of certain characteristics of the lost love object typically occurs as a sort of restitution of that object. The developing individual thus takes on the personal characteristics of a number of persons with whom he has maintained an intimate object relationship. The ego is, to some extent, a structural composite derived from an individual's past object relationships.

Identification may also imply a *defensive* process. Thus, the concept of an *identification with the aggressor* involves a process by which an individual takes on the characteristics of a threatening and more powerful external figure. Identification, when viewed in such a light, involves a form of counter-phobic mechanism by which an individual comes to grips with a threatening passive experience, by taking on the role of the active aggressor. Although Freud tended to specifically relate such a defensive process to the resolution of the oedipal crisis, it may perhaps be generalized to a far broader series of identificatory coping maneuvers. Thus, throughout childhood and young adulthood, the individual is constantly faced with external authority figures who obviously are more powerful and effective copers than he is. Identifications, however, allow for an enrichment of the individual's coping capacity through an assimilation of mastery techniques. The initially weak and incompetent individual hence becomes more masterful by internalizing the characteristics of more powerful and aggressive authority figures.

The formation of a strong ego replete with a sense of mastery and a capacity for competent coping efforts is, of course, dependent upon the *social learning* history of the individual. If the individual has been lucky enough to have been confronted with masterful and competent models then he too will take on such attributes. If, however, he has been confronted with less competent models then he will not be capable of coping effectively. The process of modelling is directly cognitive in nature and occurs by means of perceptual imitation and a vicarious sharing of internal subjective states. The perceptual salience of the model hence is an essential factor in determining his social learning impact upon the perceiving subject. Particular attributes such as physical stature and expressive style and movements take on, therefore, extreme importance. Such a social learning process is, of course, also affected by standard principles of learning, such as reward and punishment, motivational variables, and degree of practice.

The *content* of identifications range from id factors such as anger and sexual dispositions to superego factors such as moral prohibitions and value judgments. Other forms of superego identifications may include types of self-restriction, impulse control, and disciplined work habits. Ego factors such as interests, characterological formations and expressive styles are also the product of identifications. The insight provided by Schafer's ego psychological approach (1967) must also, however, be kept in mind. Schafer noted that, beyond such part processes of identification, there can also exist complex composites of identifications. Some manifest forms of behavior that outwardly appear to be determined by superego processes thus may also be integrally connected to id and ego determinants.

Although such identificatory composites and part processes are initially formed on the basis of a *dependent* interaction in which a weaker and less potent subject is required to model himself after an outwardly stronger and more competent authority figure, such is not the case once the particular traits have been adequately internalized. At a later stage, the internalized traits become *autonomous* functions integrated within the fabric of the subject's total personality constellation, and can be maintained without modelling interaction.

THE IMPORTANCE OF MODEL ATTRIBUTES

The rather broad conception developed in the present paper implies the importance of model attributes in determining the nature and extent of identification processes. A number of specific hypotheses with regard to group psychotherapy interactions are suggested by such a conception.

Hypothesis one

Physical characteristics such as the height and build of the group therapist, since they effect his "perceptual salience," should tend to affect the degree to which he is identified with by the individual group members.

Hypothesis two

Physiognomic characteristics conveyed through the therapist's habitual facial expression, bodily posture, and movement patterns should also be related to the degree of identification. Expressive styles conveying salient attitudes of confidence, activity, vibrancy, and involvement should, in general, lead to more identifications than more neutral nondescript expressive styles. No value judgment is implied by the particular expressive attributes listed above. The inference is merely being drawn that a more expressive therapist will be identified with more readily, all other things being equal, than a less expressive therapist.

Hypothesis three

Psychological factors such as the apparent competence, powerfulness, narcissistic self assuredness, aggressiveness, and potential for assertive leadership, of the therapist will effect the degree to which he is identified with by the individual group members.

Hypothesis four

Sociological characteristics such as the social status and powerfulness within a recognized social hierarchy of the group leader will effect the degree to which he is identified with.

The above four hypotheses, presented as they are in an oversimplified form are meant merely to be examples of the kind of inferences with regard to model characteristics that can be drawn from a broad conception of the identification process. They are not meant to be rigidly adhered to and are merely a series of tentative assumptions which need to be tested out through systematic observation of group interaction processes. A value judgment should not be implied with regard to the personal characteristics necessary for the *ideal* group therapist. On the other hand, it should be stressed that personal attributes of the therapist are quite relevant to whether or not identificatory interactions occur during the course of group therapy. Identification does not occur in a vacuum, nor is the therapist a mere *blank screen* upon whom the patient projects personal feelings and attitudes.

The personal characteristics of the subject are also, of course, important factors in determining whether or not an identification occurs. Thus, certain model attributes may be effective with one subject but not another. The interrelatedness between subject and model characteristics is thus also an important aspect of the identification process meriting serious systematic study.

In summary, a review of a number of different conceptions has led to an integrated formulation of the process of *identification*. The latter formulation incorporates principles derived from psychoanalysis, learning theory, social learning theory, sociology, and the psychology of cognition.

The formulation has been applied to the process of group psychotherapy, a number of hypotheses being suggested with regard to leader characteristics which should tend to enhance identificatory behavior.

References

Bandura, A. and Walters, R. H. (1963): *Social Learning and Personality Development* New York. Holt, Rinehart and Winston.

Bandura, A., Ross, D. and Ross, S. A. (1963): Imitation of Film-Mediated Aggressive Models. *J. Abnorm. Soc. Psychol.*, **66**, 3–11.

Bandura, A., Ross, D. and Ross, S. A. (1961): Transmission of Aggression through Imitations of Aggressive Models. *J. Abnorm. Soc. Psychol.*, **63**, 575–582.

Brody, M. W. and Mahoney, V. P. (1964): Introjection, Identification, and Incorporation. *Int'l J. Psychoanal.*, **45**, 57–63.

Dollard, J. and Miller, N. E. (1941): *Social Learning and Imitation*. New Haven. Yale Univer. Press.

Freud, S. (1960): *Group Psychology and the Analysis of the Ego*. New York. Bantam Books.

Freud, S. (1933): New Introductory Lectures on Psychoanalysis. New York. W. W. Norton and Co.

Freud, S. (1953): The Passing of the Oedipus Complex. In Freud, S. *Collected Papers*. Vol. II. London. Hogarth Press.
Freud, S. (1953): Mourning and Melancholia. In Freud, S. *Collected Papers*. Vol. IV. London. Hogarth Press.
Kagan, J. (1958): The Concept of Identification. *Psychol. Rev.*, 65, 296–305.
Knight, R. P. (1940): Introjection, Projection and Identification. *Psychoanal. Quarterly*, 9, 334–341.
Lazowick, L. M. (1955): On the Nature of Identification. *J. Abnorm. Soc. Psychol.*, 51, 175–183.
Mowrer, O. H. (1950): *Learning Theory and Personality Dynamics*. New York. Ronald Press.
Parsons, T. (1955): Family Structure and the Socialization of the Child. In Parsons, T. and Bales, R. F. *Family, Socialization and Interaction Process*. Glencoe, Illinois. Free Press.
Schafer, R. (1967): Ideals, Ego Ideal, and Ideal Self. In Holt, R. R. Edit. *Motives and Thought. Psychoanalytic Essays in Honor of David Rapaport*. New York. Int'l Univ. Press.
Sears, R. R. (1962): The Growth of Conscience. In Iscoe, I. and Stevenson, H. W. *Personality Development in Children*. Austin. Univ. of Texas Press.
Singer, E. (1965): Identity vs. Identification: A Thorny Psychological Issue. *Rev. of Existential Psychol. and Psychiat.* 5, 160–175.
Whiting, J. M. (1962): Resource Mediation and Learning by Identification. In Isoe, I. and Stevenson, H. W. *Personality Development in children*. Austin. Univ. of Texas Press.

Author's address: Jewish Community Services of Long Island, 22 Lawrence Avenue, Smithtown, Long Island, New York 11787.

An experiment in group consultation with the staff of a comprehensive school

A. C. R. SKYNNER

A model derived from group-analytic principles for time-limited intervention by the conjoint family method, is described in its application to another type of "natural" group, in this case to the entire staff of a large comprehensive school by a psychiatric team. Besides the surprising positive potential of this type of intervention, the special difficulties of such a situation are illustrated, including the need for adequate structure and control and for techniques which maintain relationships of mutual professional respect.

IN two previous papers, (Skynner 1964, 1968), I described a use of group situations for facilitating the growth of professional skills and development of personal insight. These followed a group-analytic model where the focus was on the developing group themes and transference behaviour, the latter usually demonstrating relationships with the typical problems of clients served by the agency to which members of the groups belonged. Such groups resemble the artificially-constituted "stranger" groups utilised in conventional small group psychotherapy, in that members are drawn from different agencies and efforts are usually made to avoid including people who are working together or likely to come into frequent contact outside. This method facilitates a free exchange between members by diminishing anxieties that revelation of professional or personal problems may have adverse effects on relationships with working colleagues, or on their reputation and standing with

---

COMPREHEND: Grasp mentally; have understanding of; be inclusive of; embrace . . . (Oxford Pocket Dictionary)

superiors, but it also creates other difficulties. There is often a "re-entry problem" in that the members of the training group work through their own resistances to new knowledge and skills, but when they return to their own agencies encounter strong resistances in others which prevents the knowledge gained from being generally utilised; indeed, those undergoing the training can often exacerbate such reactions by developing a rather precious, superior attitude which fails to allow for the natural caution of colleagues. Furthermore, even if care is taken by the leader of the training group to secure sanction for the process from those in positions of authority to its members, and to ensure that this sanction is repeatedly renewed, it is impossible to avoid arousing anxieties in those in authoritative positions that the new knowledge may cause some problems: for example, that their juniors will acquire skills they do not possess themselves, or that the situation may create divided loyalties which will undermine their own position. All those who have run such groups will be familiar with these hazards, and with the readiness of some trainees, because of unsolved personal problems, to exploit these in their work situations.

These considerations, as well as an increasing awareness of the power and potentiality of working with "natural" groups—families, or married couples—led me to believe that it might be preferable to find ways of working with agencies or institutions as a whole. Using a group-analytic model, with a focus on facilitating interaction and exchange between group members rather than on encouraging the development of transference phantasies centred mainly on the therapist, I had discovered that the main therapeutic work was carried out between sessions, (Skynner 1969). This made possible profound and lasting changes, both in family functioning and in the well-being of its members, in what seemed an impossibly short time if one considered only the few hours the family met with the therapist and neglected the time, perhaps a hundred-fold greater, during which the problems continued to be worked out at home. I hoped that agencies, institutions and other organisations in which individuals worked together as a group on a common task might be able to use a group-analytic intervention in a similar way, or perhaps even more effectively since the working group would be less likely

than families to function by means of a collusive system of defence.†
Also as in families, resistance to new understanding would be faced
and worked through simultaneously by the whole group, ensuring
that such changes as were achieved could be expected to survive
over time.

An invitation from the Islington Family Service Unit to meet
regularly with the staff as a psychiatric consultant, and later from the
East London Family Service Unit as well, gave me an opportunity
to put these ideas into practice. I met with each unit every two
months over several years, and became convinced that the possibility
of setting in train a process of growth and development in the organisation as a whole, and so in the individual members, did indeed exist
to a point where the intervention could become part of the culture
of the institution.

By 1967 I felt I had discovered an ideal method of assisting schools
with children who presented problems to them, a task which could
clearly never be met by any imaginable increase in conventional
child guidance provision, at least in the localities with which I was
concerned. At the clinic and hospital I served we had become
increasingly aware of the need to make some of our knowledge and
skill available to teachers, since so many of the cases referred to us
were unresponsive not only to child psychotherapy but also to work
with the whole family, separately or conjointly. These families were
often hostile to all social agencies, including clinic and school, or
were so inadequate and pre-occupied with simple issues of material
survival that little energy was left for other tasks. In either case,
clinic attendance was difficult to secure and co-operation rarely
prolonged even when they did come to the diagnostic interview.

---

† In fact, agencies and work organisations seem to resemble families in possessing collusive defensive systems to avoid insight and change to a much greater degree than I anticipated, not only because of emotional pressures on new members to share the agency's ways of perceiving themselves and others, but even more because of the unconscious mutual selection process—akin to that occurring in marriage—which brings together people of similar personality type or with similar psychopathology. For this reason, organisations tend to become manifestations of the personality structure of the leader, overtly mirroring his conscious beliefs but covertly demonstrating his unconscious attitudes. At least children and parents are spared the disadvantage of being able to choose each other!

Such families never came to us of their own volition but were pressured to attend because of complaints that their children showed anti-social behaviour in school; and we began to see that a better approach might lie in helping schools directly to develop more psychological understanding of these children, in order that the staff might more effectively counteract the negative influence of an anti-social home or compensate for some of the inadequacies of parents unable to fulfil ordinary responsibilities to their children. It was at this point, when we were wondering how we might best approach a school and suggest an experiment of this kind, that a request for this type of co-operation came, through the educational psychologist, from a local school.

The comprehensive school concerned was in fact ideal for such a study. It had been formed a few years earlier from a number of neighbouring schools; all were in old, dismal buildings, in an area of North East London with bad housing, marked social mobility, breakdown of traditional cultural groupings and high rates of delinquency and psychosis. Coloured immigrant children from the West Indies, often suffering the effect of two separations, (the first when their parents left them behind with relatives, the second when they left these relatives a few years later to re-join their parents), formed two-thirds of the population of the school we visited. Even in the past, before the immigrants made up such a large proportion, one of the component schools was already known as a "sink" school in which all the most difficult problems were likely to end.

Nevertheless, by the time the staff contacted us, vast efforts had been made to improve the situation. We were greatly impressed by the understanding of the headmaster and his senior colleagues, and during our visits developed a profound respect for the high morale, patience, devotion and sheer knowledge of the children's personalities and homes that the staff demonstrated. Thus, although this was a good test case for our experiment, because of the high incidence of problem children and families, we had the advantage of collaborating with teachers who were unusually co-operative, frank, and willing to learn with us. In this as in other matters, I fear that those most in need of help are usually least prepared to accept it, but it remains to be seen how much the interesting results of this experiment can be generalised.

At a preliminary meeting with the head and his deputies, held at the clinic, a basic structure for our collaboration was agreed. As it was an experiment from which we hoped to learn as much as the school, all the specialities represented in the clinic team—psychiatrist, psychologist, psychiatric social worker, and child psychotherapist—were to be represented. We would meet regularly on the school premises, with as many members of the school staff as possible, so that there might be an opportunity to facilitate a learning or growth process in the school staff as a whole. We tried hard to secure an hour-and-a-half, or at least an hour-and-a-quarter for these meetings but this was not possible. The whole staff could be brought together only by missing part of their lunch break and altering the times of classes. This gave us only an hour to play with, already dangerously short if clear conclusions were to be reached and putting us in greater peril of an inconclusive and unsatisfactory discussion if for any reason we started late. As to frequency, the head and his deputies had been thinking in terms of meeting once a term, too infrequent in my experience for any real group interaction to develop. My colleagues felt strongly that we should meet not less than fortnightly, while my own intuitive choice was for monthly meetings—frequent enough to permit the group development I hoped for, but also sufficiently spaced out to avoid arousing fears of dependency or of a "take-over" by the clinic, with all the resistance which this would arouse. In fact, when asked at our first meeting the school staff spontaneously chose monthly meetings, and this remained the rule, modified by holidays and other expediencies to about 6–8 meetings a year.

At our first meeting there was a large attendance, perhaps about 30 teachers in all. The four members of the clinic team scattered themselves around a large central table in the library where we met. Most of the teachers also gathered around this table, but some sat more peripherally at tables near the walls and this possibility of choice regarding the degree of involvement and so of a gradual approach to the situation seemed valuable; some teachers would come in and pretend at first to read a book at a distant table, later allowing themselves to be drawn into discussion. We were not at any time introduced to the staff, and there was no preamble. A deputy head, who acted as the main representative of the school on these occasions, at once took out notes and asked for advice on a child whom he proceeded to describe.

As might be expected, this first case was a hopeless one, a child already committed to an approved school so that neither they nor we could help him further. We showed them our own limitations and acknowledged that we found ourselves similarly helpless in the face of many problems of this kind. This led on to other feelings of discouragement and hopelessness in the staff about the difficulties of influencing some children through the healthy environment the school strove to provide, when the teachers were working against the influence of a whole neighbourhood with anti-social or delinquent values. They seemed to be showing us the size of the burden they had to carry, hoping that we could help yet fearing that they would feel even more inadequate by comparison if we were able to do so. We therefore sought to assist by sharing their difficulties, rather than by taking the role of experts. We recognised of course that this would lead at first to feelings of disappointment or anger that we were witholding our "magical" solutions and such responses were indeed frequent and were pointed out as they appeared.

It may have been their feeling that the large group situation was too chaotic that led the senior staff to alter arrangements for our second visit, without prior agreement or warning. Only four teachers were present, and it was explained to us that these were the only ones concerned with the case. The meeting was inevitably more formal, and the case itself centred around anxieties about control. A 12-year-old West Indian boy had aroused anxieties in the school staff by his violent rages, particularly as he had, during one episode, injured another child with a milk bottle. This boy lived alone with his grandfather, and in the course of our discussion the lack of a mother figure appeared increasingly relevant. Although the staff gradually perceived that the boy's rages were related to his deprivation and realised that he was therefore calmed by a nurturing, maternal attitude on the part of the staff, there was a constant demand that the clinic team should relieve the staff's anxiety through taking action or responsibility rather than by the more gradual and uncomfortable process of offering understanding and insight. Should not the boy be referred to the clinic, they asked? Should he be sent away to a boarding school? What would happen if he were more accurate with the next milk bottle? These insistent demands for a practical and immediate solution made real discussion difficult, and interfered with the help we could offer through our knowledge.

We emphasised the need to stick to our original plan of working with as many of the school staff as possible, but the third visit brought a crisis. We arrived on time to find the library locked, while the teachers were still eating and paid little attention to our presence in the dining-hall through which the library was approached (my colleagues began to understand why I insisted on meticulous time-keeping on our part). A teacher finally saw us and fetched the key, but the deputy head was five minutes late and made inadequate excuses for his colleagues who drifted in without apology over the next 15 minutes.

The case presented was another West Indian boy who avoided effort and difficulty, and was supported in this by an over-protective mother. She sent notes claiming that he was delicate and asking that he be excused games and exercises. The staff appeared sympathetic about the inadequacies and fears of both boy and mother and this seemed to inhibit them from acknowledging or expressing the frustion and anger they also felt. They wondered whether they should listen to his excuses. Was he really unfit for physical exercise? They appeared to seek an external authority, in the form of medical sanction, to enable them to apply the firmness which they sensed was necessary without running the risk of damaging him. The clinic team sought to help the staff see that the two needs of the child to which they were responding—to challenge the manipulation and firmly demand effort, while at the same time providing nurturance and support—were not incompatible but were in fact complementary. Throughout the discussion a difficulty in integrating these two aspects of parental care, perhaps representing a union of paternal and maternal stereotypes, appeared again and again.

Exactly as the session was due to end, a second case was raised which took us ten minutes over our time and shortened our own brief lunch period. It was no doubt significant that the problem was one of stealing!

I perceived too late the significance of this, and my colleagues were annoyed, I thought, at my allowing the school staff to steal from us. Two of them confirmed this by arriving exactly 10 minutes late to the staff conference which followed at the clinic, saying they had needed extra time to finish their lunch; but they refused to accept any suggestion on my part that their lateness had any connection with their feelings over events at the school. This refusal of my colleagues to accept an interpretation of their behaviour which seemed blatantly obvious led me to question the whole basis on which communication over such issues between professional colleagues needed to rest. I realised that the use of interpretation was really inappropriate, both between my colleagues and myself as their leader, and between ourselves and the school, because it placed those to whom the interpretation was made in a child-like, dependent role and turned them into patients, when we sought above all to share in the solution of problems from our different but complementary professional positions. At this point I also perceived that, while *interpretations about* attitudes to authority based on a parent/child model made this impossible, the *use* of actual authority within the professional structure was in no way incompatible with the maintenance of adult professional relationships based on mutual respect. In fact, the reverse was true; I had not only a right but a duty, as head of the clinic, to demand proper time-keeping at clinic conferences if everyone was to benefit from them; and our team had a similar duty to ensure that the school fulfilled their responsibilities in a situation where they looked to us for guidance. It was clear that my management of the situation, at clinic and school, was permitting anxiety to rise to excessively high levels, thereby threatening the project with breakdown and leading individuals to set up their own boundaries because those I was providing were inadequate.

It was at this point, therefore, that I became convinced that more structure was needed, and that we should accept and work within the hierarchial pattern on which ordinary schools are traditionally based, seeking to modify this temporarily only enough for our purposes. I saw that the headmaster's authority was in fact passed to us by his sanction during the session, and that I, as leader of our group, must be prepared to take his role to some extent. Accordingly I telephoned the head and "carpeted" him for his staff's lateness and the disrespect this implied. At the next session the rebuke had clearly not only been registered but also passed on down the line. The turn-out was large, everyone was on time, and the previous demoralised atmosphere was replaced by a much more keen, alert and co-operative response. I took a more active role to start with in relation to the school and my colleagues, summarising the development of the discussions until then, and linking the problems of authority in their form and content. From this point on, the discussions went well, and though the difficulties repeated later in various forms we appeared to deal with them more adequately.

The themes underlying our discussions resembled the progress of a patient in psychotherapy, or the phases through which a training group passes. First, as already described, there was a general regression as the staff brought cases where the presenting problem indicated increasingly primitive and infantile functioning in the children concerned, while the anxiety over the phantasies these cases aroused in the staff was gradually faced. This regression reached its greatest depth, I think, with the discussion of another deprived boy whose uncontrollable screaming obviously aroused panic and rejection in others. As it happened, this boy gave a blood-curdling demonstration outside the library door at a critical point in our discussion, at which I reported my own spontaneous phantasy of wishing to strangle him. This led to the expression of similarly violent but previously denied feelings in the staff, and enabled us all to accept such alarming emotions more easily and so also to accept the boy himself. Other cases discussed were usually not mentioned again for some time, presumably indicating improvement or at least increased staff tolerance, until after some months we would often be told how much improved they were. However, they felt this case was beyond them and the headmaster said firmly that the boy should be seen at the clinic. I had no alternative but to agree, but I let him wait for two months. As I hoped, he was no longer a serious problem by the time I saw him, and no further action was needed.

My notes state: "an open, warm, and friendly boy, who agreed he used to get into trouble but said he had improved because 'the teachers are different this term'. He found it hard to explain the difference but made such comments as 'if I am asked a question and I don't know it, they will tell me now, so the work is easier for me.'" A further school report obtained just before the interview stated that he was less aggressive, was working harder, was helpful to the staff, and appeared happy. There was no mention of the screaming attacks and the tone of the report was strikingly sympathetic and positive. Two months later, both mother and boy reported further improvement at home and at school and he demonstrated an even more open and friendly attitude.

After this nadir was reached we appeared to undergo a maturational process as the cases presented demonstrated successive stages of child development. The staff group, and indeed our own team, steadily became able to function with greater freedom, confidence and spontaneity. The need for the initial hierarchical structure diminished, and all members began to share responsibility more readily. Teachers seemed less threatened by the group, more confident of the value of their individual opinions and increasingly aware that honest expression of differences would lead most quickly to the truth we searched for. The developmental process in our own team paralleled or perhaps preceded that of the school. At first our own team hierarchy was more pronounced and I usually took the lead; at the first session, when I suggested we all distributed ourselves among the staff around the central table, my colleagues compromised by sitting opposite me but next to each other! Later, we functioned increasingly as a group, and if I did not feel I had a useful response to a comment by the school I could be sure someone else from the clinic would rise to the occasion. I think our own capacity to discuss and disagree openly was a crucial example for the school staff, with which they gradually identified, and they did indeed confirm this later. The development of the group is clearly seen in the following session, where sexual associations of adolescence were manifested for the first time both in the case presented and in the interaction of the staff. My report of the meeting states:

This must have been the largest gathering we ever had in the school, between 50 and 60 including ourselves altogether. The room was full, the central table crowded and all the surrounding tables occupied as well, though it was noticeable that many people could have sat centrally but preferred not to do so. People were mostly on time, seemed keen and involved, and the impression was of a very positive and interested attitude. The main feature of the case discussion was the way in which the group became split down the middle, as far as the school staff were concerned, on sexual lines. The case presented was of an adolescent boy who seemed, from many elements in his story, to be unsure of himself as a male. He lacked confidence with all the girls in his class and though he tried to win approval, they did not respect him. The staff were particularly concerned that he was known to have interfered sexually with a much younger girl on more than one occasion.

In the discussion it emerged that he caused little problem to two of the male teachers. who managed him in a way which increased his confidence, and he also seemed to be amenable with three female teachers whose relationship with him was mainly motherly and based on their perception of his feelings of inadequacy, so that they encouraged and supported him. The main difficulties arose with those teachers, both men and women, who responded to his sexual challenge as a kind of threat to themselves, and who could not see it as indicating a need for help and support.

An active, lively discussion ensued throughout the session; and much giggling and laughter with undertones of sexual excitement clearly indicated that the staff was resonating to the same sexual issues with which the boy was struggling. One of the main complainers about this boy was a very young and attractive female teacher who demonstrated a marked "masculine protest"; she seemed to feel challenged and threatened by this boy's behaviour, needing to control him and in effect castrate him. Most of the staff seemed to feel helped and satisfied by a focus on the boy's feelings of sexual inadequacy and the way in which it was revealing itself in all the symptoms; they gradually began to understand, where they had not done so already, that the boy needed help in finding a satisfactory masculine identity. We suggested the male teachers might facilitate this by offering themselves as stable, reliable, and strong paternal figures who would demand co-operation from the boy and help him to control himself, while the women could also build his confidence through helping him to feel more sexually adequate in his relationship to them, making him feel more of a man. Most people agreed with this but the pretty teacher who found such difficulty with him continued to protest. However, the school staff as a whole were obviously beginning to operate in relationship to us and each other as adolescents, talking for the first time as if they were aware of sexual differences and able to find some pleasure in them. Their amused response to their still dissatisfied colleague made us confident that they would continue to assist her maturation!

It would be misleading to leave the impression that progress in understanding and in group development was continuous. As in therapy, or in a training group, a marked advance would often be followed by a partial regression, a well attended and lively session by lateness, absence, and seemingly unrewarding interchange. The session reported above, for example, was followed by another in which many of the events of the difficult third session repeated themselves. We found ourselves once again locked out, the teachers eating; almost everyone was late, with no excuses made; and the discussions of cases seemed inconclusive and marked by indifference and fragmentation, people often talking together in two's or three's. The children discussed all showed similar characteristics, living in grandiose, wish-fulfilling phantasy, lying, stealing, and challenging the teachers' authority in subtle ways. Our psychotherapist member pointed out the denial of adolescent challenge to adult authority in these cases, and though I did not grasp it at the time, I perceived later through my colleagues' comments that the staff were saying to us, like the children to them, "we can do without you", but doing so in a way that "stole" from us by denying the value of the help we had given to make it possible. Nevertheless the situation in the group as a whole was by now very different and such challenges to us were dealt with by interchange among the staff members, as if part of this group had internalised our function.

These two sessions occurred towards the end of the second year of our meetings, when we discussed with the staff whether to continue or terminate. There was much enthusiasm for the meetings and a firm decision to continue, which our team shared, but the discussions in the following term seemed to have lost their former liveliness and vigour and attendance was less reliable. It took us some time to realise that we had all decided to continue because we had come to like and respect each other and to look forward with enjoyment to the meetings, but that our work was already done. We perceived that we had prevented the school from taking the next step of separating from us and making what they had learned their own. When we suggested this, they agreed with our assessment, and during a few more meetings, focused on working through the termination, feelings of strong ambivalence accompanied evidence that the school were already making such group discussions an integral part of their work.

DISCUSSION

Though what we learned will be apparent from the description of the experience, it may be helpful to set out the main conclusions in more compact form.

1) The visits of our team seemed without doubt a highly effective and economical way of dealing successfully with a wide range of problem children who would otherwise have been referred to the clinic, probably with limited co-operation and poor results despite a large expenditure of professional time. The whole culture of the school was changed towards a "therapeutic community" function enhancing the existing positive factors by additional psychological understanding, so that clinic involvement was necessary only for a limited time, apart of course from the on-going support through traditional clinic functions in cases they could not cope with.

2) Meeting with the whole staff avoided many problems encountered when a part of a teaching staff attends a conventional training group elsewhere. Instead of conflicts of values being stimulated between those receiving the training and those who do not, with anxieties aroused in senior staff that their authority may be undermined, everyone shared in this experience and, as with a family in conjoint treatment, such resistances and conflicts were clearly faced and were resolved at each step.

3) In working with an institution in this way, it proved desirable to accept the existing structure and relationships and to work within these, rather than attempting to impose a different frame of reference from the beginning as may be desirable in a training situation conducted away from the setting in which the actual work is carried out. Thus, in visiting schools one should be prepared at first to fall to some extent into the traditional teacher/pupil pattern; the new ethos can be communicated gradually, in the course of the discussions.

4) To preserve relationships of mutual professional respect and avoid placing the staff of the institution in a "child" or "patient" role, interpretations should be avoided in general and insights communicated by other means. Some of these are: (*a*) by focusing

on the dynamics of the case, recognising that this always reflects the current pre-occupation of the group at the time it is presented, but leaving it to the group to understand this for themselves through identification. Caplan, (1964), has of course done much to clarify this technique. (b) by demonstrating the value of freer communication through the actual functioning of the clinic team. The fact that much learning of new skills takes place through identification with the clinic team helps to justify the greater professional time involved when several clinic members participate. (c) Many issues which would be worked out in treatment by interpretation can be dealt with equally well by exchanges within the bounds of ordinary professional relationships. This is clearly illustrated by my criticising the staff's behaviour towards us in session 3 instead of interpreting this behaviour in terms of infantile conflict. (d) interpretation can nevertheless be safely used if one applies it also to oneself, thus sharing the experience and understanding rather than "talking down." My expression of my impulse to strangle the screaming boy was an example of this, allowing the staff to accept their own primitive emotions and in turn to accept the boy who had provoked them.

5) It was most important that we did not see the children discussed though the staff often asked us to do so. By working as we did we were obliged to see the children through the eyes of the staff, and they to some extent through ours. Thus, the very difficulties of communicating about the problems, provided we struggled honestly to understand, ensured that what emerged contained the psychological insight of the school as well as the clinic team and brought the two closer together to form a new and more comprehensive understanding.

6) The beneficial effects of these meetings seemed to come from this understanding, which in its turn brought acceptance and more appropriate handling. The children presented had been rejected in the sense that they had not been understood, could not be contained psychologically in the minds of those dealing with them. It was clear that these children responded dramatically to the feeling of acceptance and contact they received when the staff could look at them squarely and simply without fear or puzzlement.

7) Needless to say, some knowledge of the laws of group interaction, and of large groups in particular, is vital if one is to cope with such a complex situation. Fortunately, such training is becoming more widely available and the Institute of Group Analysis, for example, now includes a large group experience as part of its introductory Course in Group Work. Those who have worked with large groups have often prefered to do so in company with a team of colleagues, and this experience certainly emphasised the value of such an approach. A situation of this complexity is a heavy burden for a single-handed conductor, and a team is undoubtedly able to offer more adequate and flexible leadership. It was also of great importance to have the different disciplines represented. Time and again the child analyst was able to make a child's behaviour meaningful to the staff when I had no adequate answer. The Psychiatric Social Worker was able to represent the parents point of view, or that of other agencies involved, while the Educational Psychologist, who visited the school in other capacities and was already known and trusted by the staff, and who had teaching experience as part of her background, frequently played a vital role in linking us, perceiving where difficulties of communication might lie and clarifying the difficulties of each side to the other.

POSTSCRIPT

Recently, over 3 years after the experiment ended, the Educational Psychologist and I returned to take lunch with some of the staff, in order to assess the long-term effects of our meetings. There was evident pleasure on both sides at renewing our acquaintance, but a comment by one teacher that they wished they could still have our help with problems sounded more like an expression of politeness than of need. This was confirmed when, after the initial exchange of courtesies, we enquired about current difficulties. The teachers looked at one another, increasingly perplexed, and one said "whatever became of all those problems we used to meet?" The children discussed at our meetings had, with one exception, all improved, and even this exception had been contained and coped with until the normal leaving age. They found it difficult to think of any children currently presenting serious problems, and even the one they finally remembered was clearly not occasioning much anxiety.

This change had occurred in a context, according to the educational psychologist who still works in the area, of a continuing increase in maladjustment in most other schools in the locality. It was, however, impossible to decide how much our meetings had contributed to this improvement. The school had moved from its previous Victorian buildings to spacious new premises; there had been more time for the old schools combined into this comprehensive to integrate; and a new system of year-masters had been instituted to provide pastoral care. Nevertheless, other schools which had possessed similar advantages over a longer period were suffering more acute problems, and the staff spoke of the ability they now had to share problems and communicate about them, as well as the confidence they now possessed that they could cope. If they had derived some of these new strengths from their meetings with us, they had clearly forgotten their origin, so much had they made this new knowledge their own. And that, I believe, is as it should be.

References

Caplan, G. (1964): Principles of Preventive Psychiatry, Tavistock, London.
Skynner, A. C. R. (1964): Group-Analytic Themes in Training and Case Discussion Groups, in Selected Lectures, Sixth International Congress of Psychotherapy, Karger, Basle & New York.
Skynner, A. C. R. (1968): A Family of Family Case Work Agencies. *International Journal of Group Psychotherapy*, 18, 352.
Skynner, A. C. R. (1969): A Group-Analytic Approach to Conjoint Family Therapy. *Journal of Child Psychology and Psychiatry*, 10, 81. (Reprinted in Social Work Today, July 15th 1971).

Author's Address: Dept of Psychotherapy, The Maudsley Hospital, Denmark Hill, London S.E.5. England.

Terminating an open-ended therapy group

HENRY P. POWERS,
MARIA VICTORIA ACOSTA-RUA and
RICHARD P. VORNBROCK

As the therapy process in our long-term open-ended group developed to its inevitable conclusion, the authors realized that mass termination was the logical ending. To prepare ourselves for this step, we consulted the professional literature for the experiences of others.

In the literature we found plentiful material on termination of individual therapy, termination of single members from therapy groups, and planned termination of time-limited closed-ended groups. But we found no material on the mass termination of a group such as ours. While the importance of successful termination is well established and most therapists would agree generally on the goals to be achieved during termination, only a small part of the literature on the subject is of any use in preparing for the experience of terminating an open-ended therapy group.

REVIEW OF THE LITERATURE

An excellent historical perspective of the literature on termination has been given by Edelson (1963), by Schiff (1962), and by Mullan and Sangiuliano (1964). Therefore only a few words will be added here. Writing on termination began with Freud (1950) and for years literature on this topic was written by psychoanalysts in terms of individual analysis. Early writing was done from the standpoint of judging when the analysis was ready to be terminated and how the analyst should terminate.

From this unilateral viewpoint, the literature showed an evolving concern with the therapist-patient relationship, including such aspects as transference, counter-transference, and the reality feelings which emerge when both patient and therapist are faced with the fact of termination.

A new dimension was added to the literature when Bross (1959) described some of the differences between individual termination and group termination, including such aspects as multiple transference resolution, feedback on therapeutic progress from the peer group, lack of flexibility in "weaning" (e.g. the opportunity in individual therapy, unavailable in group therapy, of decreasing the frequency or length of sessions), and the feeling of leaving a family.

What is absent from the present literature is a complete collection of the responses that can be expected by the therapists from the group members during the termination process. Yet, without being fully aware of the range of potential responses, the therapists cannot adequately prepare for termination.

This article seeks to contribute to filling this gap in the existing literature. The ability of the authors to contribute toward filling the gap is derived from the termination experience of our subject group.

The experience amounted to a three-part process of planning the termination, observing the process, and finally evaluating the success of the plan in light of the observations. Through recounting this three-part process this article seeks to serve the purpose of providing a more comprehensive list of the range of responses than is presently available.

THE DECISION TO TERMINATE

Much has been written about the advantages of termination for an individual group member. There is consensual validation that he is ready to end therapy. He works through the vestiges of disabling pathology. His libidinal energies are not so much directed toward intrapsychic and interpsychic considerations of his own therapy as they are toward assisting other patients and, most importantly, toward deepening relationships in his life outside the group. The therapists and remaining members give him support and attention to these ends.

There are both advantages and disadvantages in all the members of a group going through the termination process together. Individual pacing may have to be abandoned, but there is the value of mutually experiencing similar feelings. There is less distraction from the task of ending therapy than there is for the individual member who is terminating while others proceed with ongoing therapy work. The therapists must be mindful of each member's pace and readiness, his differences and special needs; at the same time the therapists must be cognizant of the whole group and what is common to all the members.

DESCRIPTION OF THE GROUP

The research group was designed for long-term regressive-reconstructive psychotherapy for psychiatric outpatients at State Psychopathic Hospital at the University of Iowa. Sessions were held once weekly for one and one-half hours each. Group members were between 18 and 34 years of age; two were male, and four were female. The two males carried diagnoses of Chronic Undifferentiated Schizophrenia, and Schizophrenic Reaction, Paranoid Type. The four females carried diagnoses of Depressive Neurosis, Adjustment Reaction of Adolescence, Hysterical Personality, and Schizoid Personality.

The two males were original members of the group. Two of the females were in the group for 115 sessions, and two for 66 sessions. The group lasted 144 sessions. The last planned single-member termination occurred in session 118, 14 sessions before the introduction of the topic of group termination, leaving the membership at six.

When group termination was introduced, all the members were well into the middle phase of therapy and were nearly ready to terminate. All were functioning well enough for the therapists to question the need for more therapy. Any new members at this time might have disrupted the existing group cohesiveness and would have been starting therapy with a large gap between themselves and the other members; so it seemed best not to add new members.

PLANNING FOR TERMINATION

The next consideration was whether to terminate the whole group together, or to terminate each member separately until there were no members left. In order to maintain the consistent group cohesion, the therapists decided to terminate the whole group together.

Because a definite time structure would likely be less anxiety-provoking than an indeterminate plan, the therapists set a specific time limit. When the members first came into the group, the therapists asked that they give themselves 12 sessions to adjust to being in the group, so it seemed fair that they be asked to take 12 sessions to terminate.

ANTICIPATED GROUP RESPONSES

Prior to the introduction of termination, the therapists tried to anticipate the group responses to termination. For most group members we anticipated a reaction of "surprise-joy" or "shock-joy" that the therapists would indicate that the members were ready to terminate. Then there would be separation anxiety, but its extent and quality would vary from member to member and would be expressed at different times and in various ways. Later there would be expressions of anger and rejection towards the therapists for letting them go, indicating the ambivalence of "wanting to go—wanting to stay", emerging feelings which would be an aspect of transference that could not have been tackled during the "middle phase" of therapy. It was anticipated also that the members might report the return of earlier symptoms, indicating temporary periods of regression to earlier behavior or defenses. Concomitant with these anticipated reactions would be a surge of activity in therapy work, since it would become a reality that the group was not to last indefinitely. Probably there would be a gradual withdrawal of emotional investment by each member, particularly in regard to transference feelings. This withdrawal might find expression in increased discussion of reality situations and post-group plans. While there might be abortive attempts at withdrawal early in the termination process, this activity would be more expected in the final sessions.

FINDINGS

Throughout the life of the group, notes, in both process form and summary form, were kept for each session. In addition, an hour of videotape was recorded and preserved for six of the 12 termination sessions. These written records and videotapes of the last 12 sessions were reviewed by the authors to identify phenomena pertinent to termination.

The termination activity of each session was categorized under three headings: (1) Individual Member Reactions, (2) Therapist and Group Responses to Individual Member Reactions, and (3) Interpretation (of the process between the first two categories). The third heading means, for example, that if a member showed a reaction, the others would attempt to identify with the member whether or not the Individual Member Reaction pertained to termination.

The data were analyzed, comparing processes and phenomena of: (1) the multiple-member termination of the research group itself, (2) literature descriptions of both individual and group termination, and (3) the authors' other experiences with the termination process as part of psychotherapy.

Three large clusters of responses were identified: (1) Emotional-Unpleasant Reactions, (2) Emotional-Pleasant Reactions, and (3) Non-Emotional Reactions (e.g. progress reviews by patients). The categories of reactions and behavior from both the literature and our group are detailed in Table I.

Some findings were particularly striking. The frequency of emotional responses was greatest in the area of Emotional-Unpleasant Reactions. In both the group and the literature the most frequently mentioned reactions were anxiety (Edelson, 1963; MacLennan & Felsenfeld, 1968; Rubin, 1968; Zimmerman, 1968; Scheidlinger & Holden, 1966), depression (Edelson, 1963; MacLennan & Felsenfeld, 1968; Rubin, 1968; Zimmerman, 1968; Levine, 1967), anger (Edelson, 1963; MacLennan & Felsenfeld, 1968; Rubin, 1968;

TABLE I

Categories of reactions, feelings, and behavior from literature and from the research group. Italicized responses appeared only in group; brakceted ( ) responses appeared only in literature reviewed. All other responses appeared in both group and literature.

A. Emotional-Unpleasant Reactions
   1) Anxiety: Anxiety; panic, separation anxiety; *fear of future without therapy; wish to continue relationship with group members after termination.*
   2) Ambivalence: Ambivalence re: therapy process; current life situation; doubts; disbelief.
   3) Depression: Depression; sadness; worthlessness; rejection; hurt; loss; desertion; abandonment; guilt; grief; mourning; *feelings of rejection from others outside of group.*
   4) Anger: Anger; hostility; resentment; frustration; defiance; competitiveness; acting out; *negative feelings about situations outside of group.*
   5) Flight: Flight; denial; *sleepiness; fatigue; absences and tardiness from group session;* (reaction formations); *projection or displacement on others; intellectualization.*
   6) Regression: Regression; (introduction of new symptoms and stresses); exacerbation of (physical and) functional symptoms; increased dependency.
   7) Envy: Envy and jealousy re: progress and successes of other members.

B. Emotional-Pleasant Reactions
   8) Surprise: *Surprise that therapists thought members well enough to terminate.*
   9) Relief: Relief; pleasure; satisfaction; happiness.
   10) Desire to complete therapy work; intensified effort to face problems and to work on unresolved problems; effort to acquire deeper insights; interaction with therapist regarding therapist's feelings about termination.
   11) Love: Love; admiration; respect, gratification; expression of positive feelings toward therapist; gratitude; happiness; satisfaction; pleasure.

C. Non-Emotional Reactions
   12) Evaluation of process: Evaluation and review of own and others' progress; recalling positive group experiences.
   13) Weaning: Weaning; *expression of being more strong and mature;* planning for future; *self-sufficiency;* independence; projection into future; *successful handling of outside pressures; handling own problems without therapy.*
   14) Overt Expressions: Physical and verbal expressions of "Goodbyes"; (regrets for having to terminate and leave friends); final ceremony.

Scheidlinger & Holden, 1966; Levine, 1967; Fox, Nelson & Bolman, 1969; Kitchen, 1957; Moss & Moss, 1967), and denial (MacLennan & Felsenfeld, 1968; Rubin, 1968; Scheidlinger & Holden, 1966; Fox, Nelson & Bolman, 1969; Moss & Moss, 1967; Hiatt, 1965). Love (Edelson, 1963; Zimmerman, 1968; Fox, Nelson & Bolman, 1969; Vinter, 1967) was observed several times in the group and closely paralleled the citations of this response in the literature. Another reaction which frequently occurred in both the group and the literature was progress evaluation (Edelson, 1963; Scheidlinger & Holden, 1966; Levine, 1967; Hiatt, 1965; Yalom, 1970). However, while regression (Edelson, 1963; Schiff, 1962; Scheidlinger & Holden, 1966; Levine, 1967; Moss & Moss, 1967; Hiatt, 1965; Wolf & Schwartz, 1962) was often mentioned in the literature, there were few instances of this in the group. Only in the aggregate did the literature come close to identifying the full range of responses.

The recorded responses matched all but one of the anticipated responses. The authors anticipated that there might be a variation in kinds of responses according to the "phase" of termination (e.g. more Emotional-Pleasant Reactions in the final "phase") but the findings disproved this. As members became less dependent on the therapists and more aware that they themselves did the therapy work, there was less need for expression of gratitude. Only one kind of response came close to being significant by "phase" and that was the weaning (Edelson, 1963; Northen, 1969; Wolf & Schwartz, 1962) response, which was most prevalent in the middle stage of termination.

PROCESS

The following steps in the termination of group psychotherapy may be identified: (1) surprise and happiness, (2) doubts and ambivalence, (3) anxiety, (4) hostility and grief, (5) reconstruction of characteristic defense mechanisms against anxiety, (6) resolution of transference and concomitant dependence, and (7) acceptance of termination and reality-oriented adjustments.

Interestingly, one of the group members was unable to go through the above process of termination, since he could not move past the barrier of his own defense system (denial, intellectualization, projection). He needed additional help and was recommended for continuing therapy. However, through the support of the therapists and the group members, he was able to accept termination of the group without exacerbation of previous psychotic symptomatology.

SUMMARY

Termination is more than an act signifying the end of therapy; it is an integral part of the process of therapy and, if properly understood and managed, may be an important force in the instigation of change. (Yalom, 1970, p. 274).

Reactions and behavior associated with termination of therapy can be anticipated and predicted for patients, whether they are in individual therapy or group therapy. Sufficient time needs to be planned for a whole group embarking upon termination and the twelve planned sessions seemed to be quite adequate. Responses to termination were grouped into categories of Emotional-Unpleasant Reactions, Emotional-Pleasant Reactions, and Non-Emotional Reactions. Most responses were Emotional-Unpleasant Reactions; however, there may be a bias on the part of therapists to recognize and deal with unpleasant rather than pleasant reactions.

Therapists must assume the role of enablers in the termination process. They not only introduce termination but also identify and concur with member readiness to terminate. They maintain focus on the reality of termination and on reactions to the various aspects of termination. They help the members withdraw libidinal energy from the group and reinvest this energy in relationships outside the group. They recognize that, within a predictable seven-step process, each member must proceed toward termination of therapy.

References

Bross, R. B. (1959): Termination of analytically oriented psychotherapy in groups. *International Journal of Group Psychotherapy*, **9**, 326–337.

Edelson, M. (1963): *The Termination of Intensive Psychotherapy*. Springfield: Charles C. Thomas.

Fox, E. F., Nelson, M. A. and Bolman, W. M. (1969): The termination process: a neglected dimension in social work. *Social Work*, **14**, 53–63.

Freud, S. (1950): Analysis Terminable and Interminable. In J. Strachey (Ed.), *Collected Papers*, Vol. V. London: Hogarth Press.

Hiatt, H. (1965): The problem of termination of psychotherapy. *American Journal of Psychotherapy*, **19**, 607–615.

Kitchen, R. (1957): On leaving group therapy. *Psychological Newsletter*, **9**, 36–40.

Levine, B. (1967): *Fundamentals of Group Treatment*. Chicago: Whitehall Company.

MacLennan, B. W. and Felsenfeld, N. (1968): *Group Counseling and Psychotherapy with Adolescents*. New York: Columbia University Press.

Moss, S. Z. and Moss, M. S. (1967): When a caseworker leaves an agency: the impact on worker and client. *Social Casework*, **48**, 433–437.

Mullan, H. and Sangiuliano, I. (1964): *The Therapist's Contribution to the Treatment Process*. Springfield: Charles C. Thomas.

Northen, H. (1969): *Social Work With Groups*. New York: Columbia University Press.

Rubin, G. K. (1968): Termination of Casework: the student, client and field instructor. *Journal of Education For Social Work*, Spring.

Scheidlinger, S. and Holden, M. (1966): Group therapy of women with severe character disorders: the middle and final phases. *International Journal of Group Psychotherapy*, **16**, 174–189.

Schiff, S. K. (1962): Termination of therapy. *Archives of General Psychiatry*, **6**, 77–82.

Vinter, R. D. (1967): The Essential Components of Social Group Work Practice. In R. D. Vinter (Ed.), *Readings In Group Work Practice*. Ann Arbor: Campus Publishers.

Wolf, A. and Schwartz, E. (1962): *Psychoanalysis in Groups*. New York: Grune and Stratton.

Yalom, I. (1970): *The Theory and Practice of Group Psychotherapy*. New York: Basic Books.

Zimmerman, D. (1968): Notes on the reactions of a therapeutic group to termination of treatment by one of its members. *International Journal of Group Psychotherapy*, **18**, 86–94.

Author's Address: Department of Psychiatry University of Iowa College of Medicine 500 Newton Road, Iowa City, Iowa 52242

About the authors

MARIA ACOSTA-RUA is a child psychiatrist in Jacksonville, Florida.

CHARLES ARCHIBALD JR. is a Mental Health Consultant with the Indian Health Service and is founder and past president of the New Mexico Group Psychotherapy Society.

ROBERT L. BECK is a staff member at the Texas Institute of Child Psychiatry and an Instructor in Society Work at the Baylor College of Medicine.

FRANK CODY is Instructor in Psychiatry and a full-time faculty member at Southwestern Medical School at Dallas.

SIDNEY J. FIELDS is the head of psychology at the University of Arkansas Medical Center and a past president of the Southwestern Group Psychotherapy Society.

JOHN GLADFELTER is in private practice and is a Clinical Associate Professor of Psychology at Southwestern Medical School at Dallas.

MORTON KISSEN is a staff member of the Jewish Community Services of Long Island.

IRVIN A. KRAFT is Medical Director of the Texas Institute of Child Psychiatry, Associate Professor of Psychiatry and Pediatrics at Baylor College of Medicine and a past president of the Southwestern Group Psychotherapy Society.

DAVID MENDELL is a faculty member at the Baylor School of Medicine at Houston and is a leader and pioneer in Family and Group Therapy in the Southwest.

ABOUT THE AUTHORS

HENRY POWERS is a psychiatric social worker at the State Psychopathic Hospital, Iowa City, Iowa

LEWIS H. RICHMOND is a Clinical Associate Professor of Psychiatry at the University of Texas Medical School at San Antonio and is on the staff of the Community Guidance Center.

ALBERT E. RIESTER is on the faculty at the University of Texas Medical School at San Antonio and on the staff of the Community Guidance Center.

BARRY ROSSON is in residency training at the Menninger Foundation.

A. C. R. SKYNNER is senior tutor in psychotherapy to the Institute of Psychiatry, London. He is also hon. associate consultant to the Bethlem Royal and Maudsley Hospitals.

DINA LEE TANNER is a teacher and counselor in San Antonio, Texas.

RICHARD P. VORNBROCK is chief, social services, State Psychopathic Hospital, University of Iowa.

MYRON F. WEINER is Clinical Associate Professor of Psychiatry at Southwestern Medical School at Dallas, where he is active in group therapy research and teaching.

PAT WIGGINS is a psychologist at Texas Children's Hospital in Houston.